The Maturing
of
American Science:

A portrait of science in public life drawn from the
presidential addresses of the American Association
for the Advancement of Science 1920-1970.

Edited by Robert H. Kargon

|A|A|A|S|

American Association for the Advancement of Science

Washington, D. C.

1974

Designed and illustrated by Geri Lucas.
Manuscript edited by Blair Burns.
Printed in the U.S.A. by R. R. Donnelley & Sons Company.

Contents

Preface

The purpose of this volume is to bring together for the first time a selection of the presidential addresses of the American Association for the Advancement of Science. Without question, many more addresses than can be collected here deserve to be rescued from the limbo to which time and an exploding literature have consigned them. Some of these essays possess intrinsic value; others are so sensitive to their own times that they provide the general reader and the student of history a quick and painless entry into the period. The selection is my own; I bear full responsibility for the omission of many excellent and worthy addresses and for the inclusion of some that demonstrate less craftsmanship.

The addresses chosen bear upon the complex problems of science-society relationships during the past 50 years, a period in which science entered the public arena with an impact that still remains difficult to assess. My aim, of course, is to highlight important changes in science-society links during that period.

The introduction is designed to set the stage for the addresses rather than to tell their story for them. The material at the beginning of each section is intended to provide a framework for the addresses in that section. Several addresses have been substantially edited, and I have noted these. I have omitted (except in the introduction) all scholarly apparatus as secondary to my purpose; footnotes are included in the introduction only because archival research went into it and some historians may wish to assay its contents on that basis.

I ask the community of historians of American science for their indulgence; I have relied heavily upon their tuition. Especially valuable have been the insights into 19th- and 20th-century American science supplied by A. Hunter Dupree, Nathan Reingold, C. Pursell, Daniel Kevles, Charles Rosenberg, Charles Weiner, R. C. Tobey, H. Miller, and E. Mendelsohn. The comments and suggestions of N. Reingold, D. Kevles, S. Benison, D. Kubrin, and W. Applebaum have aided me immeasurably.

I owe large debts of gratitude to William Bevan and James Butler of the AAAS; Judith Goodstein, the Robert A. Millikan Library, California Institute of Technology; Robert Wood, Herbert Hoover Presidential Library; Jean St. Clair, chief archivist, National Academy of Sciences; Joan Warnow,

i

the Niels Bohr Library, American Institute of Physics; and Carolyn Fawcett, Widener Library, Harvard University.

Finally, I wish to thank Alice Quinlan of The Johns Hopkins University, who served as my research assistant. Alice's perception, energy, and enthusiasm made for me what could have been difficult and tiring into something lively and challenging.

Baltimore, 1973

Introduction

The New Era:
Science and American Individualism in the 1920's

> I believe that the historians of the future of these times will note
> as the most important thing going on in the world today, not the
> great war, but the fact of the awakening among men of science
> to the fact of the immense purpose which they have to fulfill in
> the affairs of mankind. Men of science have awakened to that;
> the people have not.
>
> —J. J. Carty (1916)

For the most part, revolutions reveal rather than alter power relationships; that is, they legitimize important changes that have been taking place during the years, even decades, before the most dramatic events. World War I provided the occasion for such a revolution in science-society relationships. The scientific community emerged from it with an activist leadership—a scientific establishment—which attempted to win for scientists and technologists new roles in society.

Academic scientists, such as George Ellery Hale, Robert A. Millikan, and Simon Flexner, joined with technical industrial leaders, such as J. J. Carty and Frank Jewett of American Telephone and Telegraph Company and Gano Dunn of J. G. White Engineering Corporation, to create or vivify institutions that would facilitate and coordinate the efforts of scientists and technologists and would impress upon the leaders of government and industry the new importance of science for national life.

Like their counterparts in financial, industrial, and governmental circles, spokesmen for science in the New Era of the 1920's espoused a philosophy of cooperative individualism, in which groups within American society would link creatively for the betterment of the nation. Scientific leaders stressed the importance of research for the continued health of the nation's industry and the proper functioning of its government in the modern world. They lobbied, therefore, for a substantial increase in material support for science, but this support was to be given within a framework of private initiative and its distribution was to be controlled by the scientific community and its leaders.

1

The Dislocations of Peace

It is perhaps by now a truism that World War I was one of those great historical discontinuities that wrenched old orders of various kinds and created new directions and opportunities in many areas of political, economic, social, and intellectual life. Of these areas, the scientific, it seems, is one of the least explored and least understood. Yet it was widely recognized, even by the time of the "preparedness" period before the war, and clearly seen during it, that scientific research and application, especially in the physical sciences, were at the threshold of vast changes.

One of the most vivid expressions of this perception was that of Robert A. Millikan, a physicist who had served the American war effort in several technical capacities. On September 26, 1919, *Science* magazine, the organ of the American Association of the Advancement of Science (AAAS), published "The New Opportunity in Science," an address Millikan had given at the University of Chicago several months before. In it, Millikan explored in some detail the "situations created by the war" and the "lessons taught by it."[1]

Although science and technology were no strangers to the conduct of war, scientists were justified in claiming that, for the first time, science had transformed the character of warfare and had become indispensable to it. Physical scientists and engineers introduced and developed such powerful tools as sound ranging (locating the position of gun batteries by computing the centers of sound waves emanating from them), submarine detection devices, accurate bomb-dropping techniques, aerial photography, aeronautic instrumentation, radiotelephony, wireless communication between airplanes, and infrared and ultraviolet signaling.[2] The meteorological section of the Army's Science and Research Division developed long-range propaganda-carrying balloons, drew more sophisticated and complete maps of the upper regions of the atmosphere to facilitate aerial navigation, and developed aids to artillery.[3] The Chemical Warfare Service greatly increased the United States' capabilities for gas warfare by developing new antipersonnel gases and gas masks. Clarence West described the Service's "crowning achievement" as the introduction of mustard gas, a blistering gas that can cause casualties up to ten days after its shell has been fired.[4] The Ordnance Department, under astronomer F. R. Moulton, created new methods for computing projectile trajectories and better tables to correct for "ballistic wind."[5]

The many impressive applications of scientific talent to war led Millikan to assert, rather extravagantly, that "for the first time in history the world has been waked up by the war to an appreciation of what science can do."[6]

> But just now [Millikan continued] the War has taught young soldiers that they need their science for success. Administrative positions in the industries are to-day being filled as never before from the ranks of the technically trained men. The War has taught the prospective officer that he can not hope for

promotion unless he has scientific training. The War has taught
the manufacturer that he can not hope to keep in the lead of his
industry save through the brains of a research group, which
alone can keep him in the forefront of progress. As a result of
all this there is indeed a new opportunity in every phase and
branch of science.[7]

This new opportunity merely reflected changes in society that had
slowly been taking place since the Civil War. The World War had caused
these changes—in government, industry, and even academic life—to reveal
themselves in an amazingly sharp, clear light and had even quickened them.
The expanding importance within American society of trained specialists
and the increasing power and prestige of science itself were now widely
recognized.[8] Thus, if the World War did not call into being America's
"technostructure,"[9] or even alter the direction of its course, it did in fact
focus attention upon the expanded role of scientists in society, and it
imbued scientists with a new confidence and a sense of a destiny soon to
be realized. Groups of well-trained scientific men were, after the fashion
of such research groups in the war effort, being mobilized in industry.
"The Ph.D. in physics, if he is a man of ability, is in demand today in the
industries as he has never been before."[10] Experimental work, for the
first time in an important way, was opening to team effort. The day of
the isolated experimenter was near an end; "[t]he war has demonstrated the
immense advantage of co-operation."[11] A young American seeking a vocation must be encouraged to consider science:

> [T]he scientist is, in the broad sense, a creator of wealth as truly
> as is the man whose attention is focused on the application of
> science. Indeed, the scientist is merely the scout, the explorer
> who is sent on ahead to discover and open up new leads to
> nature's gold.[12]

Millikan also saw a new opportunity for those who already possessed
wealth to invest in the scientific enterprise, for no better investment
existed. The United States, Millikan went on, has not been the leading
scientific nation—not because of lack of natural abilities nor even of
facilities, but rather because of the lack of determination of superior young
men to choose science as a career.

> Our greatest need is not for more facilities, but for the selection
> and development of men of outstanding ability in science. Find
> a way to select and develop men, and results will take care of
> themselves.[13]

It was not through government initiative that this development would
come about; such socialism was anathema to Millikan. In America, private
initiative may be summoned to the task of creating research chairs and
centers:

> Most of our great advances in the past have been through
> private initiative, and I suspect Mr. Elihu Root was, as usual, a
> wise counselor when he said recently, in substance, "If we are
> going to conserve the finest elements in Anglo-Saxon
> civilization, we must conserve the method of the private
> initiative and not depend primarily upon government aid." [14]

The lessons the War taught science—the importance of scientific research for the nation's economy and security and the importance of cooperative efforts for the scientific enterprise itself—were grafted upon a legacy of 19th-century liberalism. The result was a paradox. Whereas scientists like Millikan and his colleague Hale doubtless felt the lure of government intervention, with all the resources and power it would bring, they likewise feared government control or interference and remained steadfast in their faith in the altruism of American business leaders and the ultimate social wisdom of development along lines laid down by American individualism.[15]

Hale, an astrophysicist and statesman of science, had concerned himself with the advancement of science in the United States, and with the National Academy of Sciences' role in its progress, ever since his election to that body in 1902. In a series of articles that appeared in *Science* from 1913 to 1915, Hale pushed for expansion of the Academy's role in American science, especially in the custody and disbursement of funds collected in trust for scientific research. "It is safe to predict [Hale wrote] that the privilege of securing the Academy's aid in the control and disbursement of large sums for the benefit of science will be widely sought in the future." [16]

The National Academy of Sciences was founded in 1863, by an act of Congress, for the express purpose of providing scientific advice to government agencies requesting it. In the years between the Civil War and World War I, the Academy was called upon but little. Membership was an acknowledgement of prestige but did not confer access to power. Hale, however, saw in the Academy a body around which scientific research interests of all types—industrial and academic, as well as governmental— could be organized. Yet, like most members, he feared the political ties that might bind and damage its independence. Much of the ensuing history of the Academy's science politics is the story of Hale's skillful attempts to steer a course between the Scylla of government interference in science and the Charybdis of the scientific community's traditional lack of money and power.

The crisis surrounding the sinking of the *Lusitania* gave Hale his opportunity to use the Academy to attain his goals and at the same time increased the risk of closer links with the government. In a day letter dated July 13, 1915, Hale told Academy president William H. Welch, of The Johns Hopkins University, that "Further reflection and conference . . . convinces me that the Academy is under strong obligations to offer services to the President in the event of war with Mexico or Germany." [17] Welch was reticent. He believed that Hale's plan had to be considered by the Academy's Council, a somewhat involved task in midsummer. "I can imagine," he replied, "no objection to the Academy offering its services to the President in the event of war—I think an improbable event while Wilson is president—but I am not clear about our organizing a committee at present."[18]

The urgency of Hale's concern may have stemmed from his fear that the Academy would be eclipsed as technical advisor to the government by Thomas A. Edison's Naval Consulting Board. In July 1915, Secretary of the Navy Josephus Daniels had asked Edison to head a board concerned with organizing American inventiveness with regard to naval warfare, especially with regard to new possibilities such as submarine attack. On July 13, Edison's personal representative advised Daniels of Edison's acceptance, and the Board, completely independent of the National Academy of Sciences, was constituted.[19] Regarding his own plans, Hale communicated to Welch his view that "to disarm possible criticism and assure success we should concentrate on medicine and surgery but include representatives of other subjects on the committee. Daniels' announcement today regarding Edison confirms this."[20] After the March 1916 sinking of the *Sussex* and President Wilson's April ultimatum to Germany, the National Academy of Sciences, at the urging of Hale, officially offered its services to the nation. The immediate result was the formation of the National Research Council (NRC), with representatives from industry, education, scientific societies, and government agencies. In some sense a child of preparedness, the NRC quickly became an effective body when war began.[21]

The coming of the war and the preparedness theme thus provided the occasion for, but not the source of, the established scientists' concern with systematizing research and its relations with other institutions of society. By the end of 1913, the AAAS had organized the Committee of One Hundred to promote scientific research in America. In a meeting at the Cosmos Club in Washington, D.C., in April 1914, leading members of the AAAS, including A. A. Noyes, Hale, J. McKeen Cattell, E. C. Pickering, and Ira Remsen, discussed such questions as the feasibility of a central bureau for research under the AAAS, the Academy, or the Smithsonian Institution; research in industry and educational institutions; and the selection and preparation of scientific men.[22] Committees were formed, meetings held annually for several years, and reports published. One of the most impressive of these reports was C. K. Mees' "Organization of Industrial Research Laboratories," in which he argued on behalf of the Committee of One Hundred that "for industries to retain their position and make progress they must earnestly devote time and money to the investigation of the fundamental theory underlying the subject in which they are interested."[23]

The Committee rapidly became obsolescent as the war progressed and as the National Research Council grew in importance.[24] Indeed, the mainstays of the NRC—including Millikan, Carty, J. C. Merriam, Hale, Noyes, R. Yerkes, J. Angell, and C. D. Walcott—were also at the heart of the Committee of One Hundred. In any case, the National Academy of Sciences and its daughter, the NRC, increasingly held center stage.

It was, in fact, the NRC that claimed the leading role in advancing the interests of scientific research in peacetime America. Under the effective leadership of Hale, Millikan, Noyes, Carty, Dunn, and others allied with them, the NRC pressed for an executive order to make the body a permanent part of the Academy. President Wilson signed such an order in May 1918.[25]

By mid-1919, the NRC, in carrying out its aims of fostering co-operation in research and stimulating both pure and applied science, had begun to forge links with major corporations and private foundations. The NRC's industrial advisory committee included such men as T. N. Vail, George Eastman, Cleveland Dodge, Pierre S. DuPont, E. H. Gary, A. W. Mellon, E. W. Rice, and Elihu Root; the NRC received large grants from the Carnegie and Rockefeller foundations.[26] Some of the aims of the NRC establishment are best summed up in their 1918 draft of a "Proposal for the Endowment of Research in Physics and Chemistry." Citing industrial competition from abroad, the draft noted that pure research in the United States was absolutely essential for industry and, significantly, "must be financed from private funds." Stressing inadequate manpower in scientific research and insufficient training, the draft recommended the establishment of three or more major laboratories, on the East and West coasts and in the Midwest, and research fellowships for which support was forthcoming from the Rockefeller Foundation.[27] The plans of Hale and the others were limited by the NRC's dependence upon private funds, and, although the NRC was effective in its reduced sphere, many of the early hopes remained far off.[28] The NRC and the Academy did, however, provide a platform for the Hale-Millikan-Dunn-Carty group, which remained visible and vocal. "We all agree," Hale wrote to Merriam, "that the only real chance of making the Research Council a complete success lies in placing the control of its various activities in the hands of a group of very strong men. The possibility of securing new funds emphasizes this necessity." [29]

This "group of very strong men" around Hale was not, however, without its opposition. Cattell, from a position within the science establishment, as a member of the Academy, favored government support of scientific research, ostensibly a more democratic approach to science policy than private funding. "You [Cattell wrote to Hale in 1920] believe that aristocracy and patronage are favorable to science; I believe that they must be discarded for the cruder but more vigorous ways of democracy." Hale's retort, very pointed, claimed expediency and charged hypocrisy: "I . . . believe in adopting what appears to me to be the most promising means of advancing science and research under existing conditions. If I am not mistaken, you would not object to an appropriation by the

Rockefeller Foundation toward the support of work in which you are interested."[30]

The ambivalence of the NRC group concerning federal intervention in research was brought into the open by two congressional bills to support research by setting up engineering experiment stations—the so-called Newlands and Smith-Howard bills.[31] At the first organizational meeting of the NRC, held on September 20, 1916 and attended by Carty, Dunn, Welch, Flexner, Hale, Millikan, Noyes, M. I. Pupin, and W. R. Whitney, among others, the issue of federal aid was broached. Hale was unanimously elected chairman, and Millikan was asked to report on the Newlands Bill. The bill provided for a $15,000 annual appropriation to each of the states "to be applied to research and under the direction of the land-grant college." The Academy formed a committee, headed by Millikan, to advise Senator Newlands on possible modifications. The committee urged the establishment in each state of a research board, to consist of the governor of the state, a nominee of the Department of Commerce, and two nominees of the NRC, thus assuring *de facto* NRC control.[32] In November, a joint conference of representatives of land-grant colleges, state universities, and the NRC was convened, but little was settled. Carty, speaking on behalf of the NRC, played down the importance of federal aid:

> [T]he importance of this Federal aid does not loom up so large in my mind as in others. Get the people stirred up: train your research men; I can't get enough of them; I want men with the right spirit. I think we are laying too much stress on the Federal aid and not enough on getting the big idea started, and we must not weaken our position by getting into any controversy or taking sides. Unless we get this public sentiment started up right Federal aid won't help us very much.[33]

In principle, the NRC group that Carty represented was prepared to accept federal aid to the states, but each bill was found to be deficient in some way. At best, their commitment was lukewarm.

Meanwhile the Newlands Bill was succeeded in 1917 by the Smith-Howard Bill (the draft legislation was written by P. V. Stephens) and the Gronna Bill, both intended to aid industrial and engineering research. Both the Newlands and Smith-Howard bills espoused the notion of the experiment station. Whereas the Gronna Bill would have set up a rather complicated system of state boards appointed by a national board, the Newlands Bill located research facilities at the land-grant colleges. The Smith-Howard proposal provided for facilities at an engineering school or university chosen by the state legislature; the Gronna version vaguely left the matter in the lap of such laboratories and individuals as were available. Each bill gave the states and territories $15,000 per annum, but the early versions of the Smith-Howard Bill gave $25,000 to the National Bureau of Standards, which would have served as a coordinating body.[34]

The executive committee of the NRC met on April 4, 1917, and voted to express its lack of support for Smith-Howard. Stephens was so notified;

he was also informed that the NRC had "no intention of discussing it further."[35]

But Smith-Howard was resilient. In 1918, a national committee for the support of the Smith-Howard Bill was formed. P. G. Nutting, of Westinghouse Research Laboratories, claimed that the bill "will do more for our national progress and welfare than would any other single act of Congress."[36] Hale, however, termed the proposed role of the National Bureau of Standards "a serious blow to science in this country."[37] He preferred engineering research to be left to the NRC.[38]

In January 1919, a national conference on the Smith-Howard Bill was called by Richard C. Maclaurin, president of the Massachusetts Institute of Technology (MIT). Representatives of various factions supporting and opposing the bill were present; Dunn represented the NRC. A report of the meeting indicates that "agreement as to the desirability of securing Federal aid was practically the only tangible result of the conference."[39] A memorandum of February 1919 lays out the fears of the NRC:

> "Federal support for research" is highly desirable in any
> institution . . . but it should be applied only according to broad
> and thoroughly conceived principles so as to produce the best
> results. . . . The establishment of an engineering experiment
> station in every Land Grant college is no guarantee that one
> college will either find the man or produce the ideas and
> practical results aimed at in the bill. Class (a) institutions will
> be more hampered than helped if called upon to render routine
> public service.[40]

The NRC, including Millikan, appears to have favored drastic modifications that would set up a research board of five scientists and engineers in each state and direct this board to pass on project proposals.[41] None of these bills became law.

The NRC emerged in the postwar period without serious challengers, but the problem of defining its own role and finding resources remained. It was at this point that the NRC's links with industry and major foundations had both rationale and utility. Opponents of the Hale-Millikan group saw the situation clearly. Cattell, for example, remarked bitterly in 1922:

> Whether the Research Council belongs to the National
> Academy, or the National Academy belongs to the Research
> Council, or both are satellites of Pasadena [Hale, Noyes, and
> Millikan were at the California Institute of Technology,
> Pasadena] is a problem of three bodies that is difficult of
> solution. The Carnegie Corporation, the Rockefeller Foundation
> and the National Research Council are another problem of three
> bodies. The Research Council depends on the endowed
> establishments for support, but the chairman of the Research
> Council became president of the Carnegie Corporation and its
> secretary has become a trustee of the Rockefeller Foundation.

> Scientific research certainly needs all the money it can get; [but]
> it may in the long run be safer and even more profitable for
> men of science to be free from the charity and control of the
> classes of privilege and sell their services to the people for what
> they are worth.[42]

Despite such criticism, the NRC group had few doubts about the wisdom of their chosen path. Millikan was one of the most articulate exponents of their philosophy. In the same year (1922) that Cattell decried reliance upon the charity of the privileged classes, Millikan warned of excessive reliance upon government:

> [O]ne of the most dangerous tendencies which confronts
> America today is the apparently growing tendency of her people
> to get into the habit of calling upon the state to meet all their
> wants. The genius of the Anglo-Saxon race has in the past lain
> in the development of individual initiative.[43]

Indeed, the progress of the race has been secured by private institutions supported by "men who have treated their wealth and their ideals as public trusts." The progress of the nation and the future of science are intertwined with the spiritual values that have infused them both:

> [T]he salvation of the world [Millikan continued] is to be found
> in the cultivation of science together with the cultivation of a
> belief in the reality of moral and spiritual values.[44]

In an address delivered at Stanford University in June 1923, Millikan even more pointedly stated his view on the future of research support in the United States:

> [P]ublic spirited men are going to see more and more that the
> support in a large way of scientific research is an investment
> which brings the largest returns of satisfaction to themselves
> and of progress to mankind which can be made at all.[45]

The growth and prosperity of the nation depend upon scientific research; efforts to redistribute wealth will be "utterly trivial in comparison with those which may be brought about by physical and biological research."[46] Having chosen philanthropy as the support of scientific research, the NRC group would in the years to follow exhort businessmen and financiers to come to the aid of science as a duty to American individualism and as an investment. Cattell's prescient warning that "We should no more depend on . . . private philanthropy for this than for ships of war" went unheeded.[47] Upon receiving the Edison Medal in 1924, Millikan asked rhetorically, "Why is it that 'fifty years of Europe is better than a cycle of Cathay'? Is it not simply because in certain sections of the world, primarily those inhabited by the Nordic race, a certain set of ideas have got a start in men's minds, the ideas of progress and responsibility?"[48]

The intertwined themes of science, responsibility, and progress would be of major concern to the NRC group in the 1920's, for it was the business-financial community to which they turned in an era that belonged to "businessmen." The message was summed up later by Millikan, when he addressed the New York Chamber of Commerce: "You may apply any blood test you wish and you will at once establish the relationship. *Pure science begat modern industry.*" [49]

The Growth of Industrial Research

It is no surprise that a major pillar of the NRC's postwar program was the relation of pure science to modern industry through industrial research. The NRC industrial advisory committee provided a direct link between the scientists and the leaders of industry. Its work began in cooperation with the Engineering Foundation in May 1918. The original steering group included A. D. Flinn of the Engineering Foundation, G. K. Burgess of the National Bureau of Standards, Mees of Eastman Kodak, and Whitney of General Electric.[50] By 1919, the advisory committee was composed of major industrialists and financiers. Vail (the president of American Telephone and Telegraph who chose Carty as his chief engineer [51]) was chairman, and others included Dodge, Eastman, Mellon, Gary, DuPont, and Root.[52] In June 1919, the NRC organized what came to be called the division of research extension for the general purpose of stimulating "industrial administrators to broaden their research activities." [53]

The major industries had already demonstrated their concern for research and, especially in the larger aggregates, had established important research centers. As early as the 1880's companies such as Standard Oil Company of New Jersey were capitalizing on the discoveries of such chemists as Herman Frasch and, later, William Burton. Eastman Kodak Company (with Mees), American Telephone and Telegraph, Parke, Davis and Company, General Electric Company (with Charles Steinmetz, Whitney, and Irving Langmuir), and E. I. DuPont de Nemours and Company all had by World War I flourishing departments or divisions devoted to research in their respective fields.[54] In 1901, Rice reported to General Electric stockholders that "it has been deemed wise during the past year to establish a laboratory devoted to original research with the aim of opening up new and profitable fields." [55] By 1910, American Telephone and Telegraph's Jewett, who had taken his doctorate in physics at the University of Chicago and was a close friend of Millikan, was seeking Ph.D.'s in physics who had studied with Millikan for work on the telephone repeater.[56] Still, the number of research laboratories remained small, and many thought the importance of industrial research to the nation was insufficiently recognized.[57]

The situation began to alter radically with the outbreak of World War I. Jewett wrote in his essay "Industrial Research" for the NRC in 1918:

> Newspapers, magazines and periodicals are continually
> publishing articles on it; vast numbers of people are talking . . .
> about it; and industries and governmental departments . . . are
> embarking or endeavoring to embark upon the most elaborate
> research projects.[58]

Before the American Institute of Electrical Engineers, Jewett claimed that "[a] survey of present day conditions in the United States discloses the greatest progress in and the greatest results from industrial research. . . . So effectively has [the] dependence on research methods been established in the larger electrical manufacturing organizations that no question is ever raised as to the proper method of attacking a problem or a difficulty." [59] Jewett may have been exaggerating, but there is still no doubt about the ability of industrial research to command national attention.

Among the most prominent of the major industrial research laboratories were those of American Telephone and Telegraph (including those of its subsidiary Western Electric Company), General Electric, Eastman Kodak, and Westinghouse Electric and Manufacturing Company. In 1907, the three laboratories of the Bell Telephone Company were merged as the engineering department of Western Electric. Carty became chief engineer and director of research; he was assisted after 1912 by Jewett, who succeeded him as chief engineer in 1916 and who later became a vice-president of American Telephone and Telegraph, as well as president of Bell Laboratories, in the 1920's.[60] Jewett was particularly keen on recruiting Ph.D.'s for the research laboratory, and a relatively large number (at least ten) joined the Bell system before 1915, including Oliver Buckley, H. D. Arnold, and, later on, H. E. Ives, Karl Darrow, and Clinton Davisson. Under Carty, research was intended to be narrowly focused toward practical, industrial ends; basic research, under the prevailing notions of division of research labor, was to be left to the universities.[61] This restriction became increasingly untenable.

General Electric's laboratory, on the other hand, had maintained a strong interest in basic research from the very beginning of its existence. Supported by Steinmetz, technical director E. W. Rice planned both basic and applied research and raided MIT for Whitney and W. D. Coolidge, his first two directors.[62] The most impressive original research at General Electric was carried on by Langmuir, a student of Walther Nernst who joined the company in 1909 and was encouraged by Whitney in his work on the physics and chemistry of vacuum tubes. Shortly after the General Electric laboratory was established, C. E. Skinner founded Westinghouse's research department in East Pittsburgh. Apparently, Skinner's balance between short- and long-term profitable research was unsatisfactory to the firm, and he was ultimately asked to resign.[63]

On the whole, however, both industry and the public were awakening to the promise of industrial scientific research—"A Scientific Santa" proclaimed the New York *Times*—as more and more firms moved toward systematized research and development.[64] Still, a precise role for industrial research within the larger scientific community had not yet emerged. In 1916, Carty extolled the great awakening of America to the possibilities of industrial research. "In the present state of the world's development," he said before the American Institute of Electrical Engineers, "there is nothing which can do more to advance American industries than the adoption by our manufacturers generally of industrial research conducted on scientific principles." [65] A great danger, he continued, remains that we may, in our appreciation of applied research, neglect the pure science upon which applied science rests. The distinction between pure and applied science is one of motive, not methods.

> By all who study the subject it will be found that while the discoveries of the pure scientist are of the greatest importance to the higher interests of mankind, their practical benefits, though certain, are usually indirect, intangible, or remote. Pure scientific research unlike industrial scientific research, cannot support itself by direct pecuniary returns from its discoveries.[66]

That industry might replace the university as the major theater of scientific research, Carty dismissed as "unworthy." Carty envisioned a division of labor: the university to provide basic scientific research and trained researchers, the industries to provide the application of that research and employment. Where, however, is the money for basic science to come from?

> It should come from those generous and public-spirited men and women who desire to dispose of their wealth in a manner well calculated to advance the welfare of mankind and it should come from the industries themselves, which owe such a heavy debt to science.[67]

It was, after all, philanthropy and enlightened self-interest on the part of industry which were expected to provide.

Steinmetz, of General Electric, likewise saw the desirability of a division of scientific labor, but he defined it in a different way. It was not necessarily motive that separated university and industrial research: "there is no scientific investigation, however remote from industrial requirements, which might not possibly lead to industrially useful development"; that is, "any scientific research whatsoever thus is industrially justified." [68]

As an example, Steinmetz pointed to work undertaken at General Electric on dielectric phenomena in air. No immediate industrial use was foreseen, but the work was justified, quite properly, on the grounds that knowledge of such phenomena might increase the industrial demand for long-distance power transmission apparatus. Not long after the research was begun, it led to a major redesign of almost all high-voltage transmission apparatus.[69] Where a division of labor is justified, however, is in the use of facilities. Some research work can best be carried out by university laboratories, other work requires the specialized equipment available only in industry.

> In general, for industrial research better facilities are available,
> but high class skilled labor, of investigators and research men
> such as available in university research by the graduate students
> is expensive in the industry. Thus researches requiring little in
> facilities but a large amount of time and attention of research
> men are especially adapted to educational laboratories, while
> investigations requiring large amounts of material or of power
> rather than time of the investigators are specifically adapted to
> the industry. . . . Efficiency thus should require a division of
> research between educational and industrial laboratories.[70]

Implicit in Steinmetz' statement is the perception that the pure-applied
science relationship was more complex than one might at first think—
that is, a sharp division of motive, which Carty maintained, is untenable.
Steinmetz seemed to see, or at least to admit, that the universities provide
a source of inexpensive labor for the industries that draw upon them. It
is interesting to note that science planners in Europe during the 1920's,
particularly those in Germany, Great Britain, and the Soviet Union, were
seriously weighing the passing of old research organizations and the need
for integrated research institutes; they saw in the United States such
prototypes in the major industrial research laboratories.[71]

In the United States, however, the sharp division between industrial
and university research went, for the most part, unchallenged. As early
as 1917, Jewett and Skinner spoke on the subject before the Philadelphia
meeting of the American Institute of Electrical Engineers. Skinner acknowledged how difficult it is "sharply to maintain . . . a clear-cut distinction
between pure and applied science," at least regarding methods, but he
proceeded nevertheless to map out the division of labor among industrial,
educational, and governmental enterprises.[72] Jewett also acknowledged
Carty's lead and insisted:

> On the one hand, the academic institution, with its pure research
> laboratories, is engaged primarily in the education and
> development of men trained in scientific methods and in
> extending the realms of knowledge for the benefit of mankind
> and without any particular utilitarian motive. On the other
> hand, the industrial research organization is engaged primarily
> in the utilization of trained scientists for the solution of
> utilitarian problems; in other words, in the cultivation of the
> regions discovered and mapped out by the experimenters in
> pure science.[73]

The point hammered home repeatedly by such spokesmen as Carty, Jewett,
and Skinner was the importance of safeguarding the province of university
research. All acknowledged and stressed the debt owed to pure research
for scientific principles and for trained manpower. All seemed confident
that by 1900 the day of industrial research had arrived and were concerned
lest its wellspring dry up for lack of support. These themes were stressed
repeatedly by Carty in the 1920's, as a spokesman for the NRC.

> In order to encourage those engaged in the industries . . . and in
> commerce to make contributions to the support of scientific
> discovery in the universities and other institutions, and more
> particularly in order to justify them from a business standpoint
> in so doing, it is necessary [Carty claimed in 1920] to
> demonstrate the pecuniary value of science. I have endeavored
> to combat the unappreciative views so often held concerning
> pure science in the universities, and at the same time I have
> urged the great practical usefulness and profit to be derived from
> scientific research conducted within the industries.[74]

In a sense, however, the prophecies of Carty and the others concerning
the increasing importance of research for industry were self-fulfilling.
Through their own direct effort, and by the force of example, the leaders
of post-World War I industrial research stimulated its spread and growth.
In 1920, a survey conducted by the NRC listed about 300 laboratories
doing industrial research; by mid-1921, the number had grown to 526,
and by 1928 to the stupendous number of 1,000.[75] The number of corporate
units initiating research peaked sharply around 1920–21, 1927–28, and
1930–31.[76] The success of what was termed the industrial research movement often, however, tended to erase in fact those distinctions which its
leaders maintained in principle. Corporations in rapidly expanding, science-
based industries were increasingly seeking out innovations to gain competitive advantage; in such fields as electronics, for example, innovations
demanded exploration of hitherto uncharted areas of physics. The distinctions between pure and applied science were being erased; the neat
division of labor suggested by Carty was obsolescent even in his own
laboratories. Nobel Laureates emerged from Bell Laboratories (Davisson)
and General Electric (Langmuir). The work of Davisson and L. H. Germer
on electron diffraction and the electron emission studies at Bell Laboratories proved important for "basic" research as well as for "applied" work
on vacuum tubes. Researchers at DuPont and Standard Oil made substantial advances with their polymerization studies; Westinghouse scientists
provided fundamental work on discharge phenomena in gases; Eastman
Kodak's K. C. D. Hickman investigated phenomena of high vacuums.[77]
Arnold, of Bell Laboratories, conceded in 1926 the erasing of the lines
of demarcation between pure and applied science:

> Our research problems are scattered along the whole frontier of
> the sciences which contribute to our interests, and extend
> through the fields of physical and organic chemistry, of
> metallurgy, magnetism, electrical conduction, radiation,
> electronics, acoustics, phonetics, optics, mathematics and even
> of physiology, psychology, and meteorology.[78]

"Science Serves and is Served by Industry" the New York *Times* noted
with ardent approval.[79]

Science and Government in the New Era

A second pillar of the NRC's program was the research establishment of the United States government itself.[80] The governmental division of the NRC, under Walcott, was organized at the end of 1919 and included an executive committee representing the departments of Commerce, Agriculture, War, and the Navy.[81] By the early 1920's, it had evolved into the division of federal relations and had enlarged to include representatives from the departments of State, Justice, Interior, the Treasury, and Labor, as well as the Smithsonian Institution. Research-related agencies were represented by their directors and chiefs: C. F. Marvin (Weather Bureau), K. F. Kellerman (Plant Industry), C. L. Alsberg (Chemistry), E. W. Nelson (Biological Survey), F. G. Cottrell (Mines), S. W. Stratton (Standards), H. M. Smith (Fisheries), W. Bowie (Geodesy, Coast and Geodetic Survey), and G. O. Smith (Geological Survey).[82] The above list indicates a healthy growth in federal scientific work. The Progressive movement had strengthened the bureaus concerned with conservation, and in 1901 the National Bureau of Standards was created, with authority to commit resources for research on standards, physical constants, and so forth for the government and for private corporations and societies.[83] The postwar boom in the creation and expansion of science-related industries put new demands on government science, and on the Bureau of Standards in particular. Postwar inflation and the economic difficulties of 1920 unleashed congressional disapproval of "inflated" government expenditures.[84] Edward B. Rosa, chief physicist of the National Bureau of Standards, published a series of articles in the *Scientific Monthly* (and elsewhere) defending government expenditure for research.

Rosa's work is of unusual significance, not only for its important conclusions and recommendations, but also for the picture it presents of federal research. Rosa provides a review of government research, education, and development concerns and data to show that only about one percent of federal appropriations were for these concerns—a figure, according to the author, which is much too low. Rosa's argument is summed up as follows:

> One per cent of the total expenses of the government spent in this constructive way seems a very small proportion in view of the wide range and the economic value of such work. . . . The government would spend less for its purchases if it spent more in standardizing the products purchased and in testing deliveries systematically . . . research and development work by the government develop wealth and the burden of taxation is thereby lightened. . . . Intelligent research by the government, in cooperation with the industries, is like seed and fertilizer to a farmer. It stimulates production and increases wealth, and pays for itself manyfold. It is as productive and profitable in peace as in war.[85]

Rosa's views on the value of research and development for national efficiency and economic growth conformed quite closely to those of his new boss, Herbert Clark Hoover, named Secretary of Commerce in 1921.

Hoover's career as a successful engineer and businessman and his work for humanitarian causes during the war had won him national attention. Vitally concerned with postwar economic recovery, Hoover launched in 1920 a campaign against waste and for national efficiency. As president of the Federated American Engineering Societies, Hoover appointed the Committee on Elimination of Waste in Industry, a group of engineers dedicated to bringing the principles of engineering to bear upon the problem of efficiency.[86] Hoover addressed the engineers early in 1920, exhorting them to apply their expertise to the problems of waste, hunger, and social upheaval:

> The kind of problems that present themselves are more
> predominantly economic . . . than at any period in our history.
> They require quantitative and prospective thinking and a sense
> of organization. These are the sort of problems that your
> profession deals with as its daily toil. You have an obligation
> to continue the fine service you have initiated and to give
> it your united skill.[87]

Science featured the new Secretary of Commerce in its April 29, 1921, issue. On the subject of reorganizing the federal government for greater efficiency:

> We want no paternalism in government. We do need in
> government aid to business in a collective sense. . . . We need
> a department [of commerce] . . . that can give scientific advice
> and assistance and stability to industry in furnishing it with
> prompt and accurate data.[88]

Hoover, "the Great Engineer," was already widely recognized as an ally of the scientific community. A letter that Hale wrote Hoover in March 1921 implies that the chairmanship of the NRC had been offered Hoover:

> I wish it might have been feasible for you to accept the
> Chairmanship of the National Research Council, as I recognize
> how tremendously your broad conceptions and great influence
> would have developed its work. As this could not be, I hope
> you will still bear the broad possibilities of the Council in mind
> and permit us to consult you regarding the best means
> of utilizing them.[89]

Hale was concerned with bringing the Secretary of Commerce into close touch with the NRC. Hoover was asked to serve, and in fact did serve, in the NRC's division of foreign relations.[90] By 1924, Hale was inviting Hoover to be principal speaker in honor of Millikan's Nobel Prize (an invitation that Hoover declined) and noted that "[I] also wish to discuss with you plans for organization this spring of National Research Fund mentioned to you two years ago." [91] When the directorship of the Bureau of Standards became vacant in 1922, upon the resignation of Stratton, Hoover turned for advice to the NRC group. Jewett recommended Millikan and Dunn for the post; he spoke somewhat deprecatingly of another leading candidate, F. K. Richtmyer of Cornell University, as "the college professor type of candidate." [92] Hoover took Jewett's advice and offered the directorship to the NRC's Dunn, who, however, declined such a radical and "permanent change of life." [93]

As Secretary of Commerce, Hoover moved quickly to implement his conception of government as the coordinator of enterprise and the supplier of information. Research and development were used to combat not only waste, but also foreign monopolies. American independence and competitiveness in world markets rested upon scientific and technological ingenuity.[94] In his *Memoirs,* Hoover claimed to have enlarged the research work of the bureaus of Standards and Fisheries and of the Radio and Aviation Division of the Department of Commerce. His friend Vernon Kellogg credited him with greatly increasing the research funds available to the scientific bureaus of the Department of Commerce.[95] A National Bureau of Standards press release extolled Hoover's view that the Bureau was "an instrument which could be used with telling effect in carrying forward his plans for the elimination of waste in industry. . . . Many hundreds of researches are continuously in process, all designed in some way to improve the quality of commodities required for common public use." [96]

Scientific research became increasingly important in Hoover's scheme of things—at least his public statements about the importance of science had increased in strength and number by the middle of the 1920's. The *Fourteenth Annual Report of the Secretary of Commerce* (1926) for the first time devoted a special section to scientific research: "The improvement of some machine or process is of great value to the world but the discovery of a law of nature, applicable in thousands of instances and forming a permanent and ever-available addition to knowledge, is a far greater advance." [97]

The NRC group had secured their ally, a fact of which they were obviously aware. During Hoover's campaign for the presidency in 1928, Hale and Millikan were enthusiastic supporters. Hale was a contributor to the campaign fund and predicted that Hoover would provide "an administration such as no other man could give this country"; "I strongly favor [Hoover] for President." [98] After the election, Hale gleefully insisted that Hoover's presidency opened a new era in American progress.[99] Millikan would continue to see in Hoover, even by 1932, a leader employing "the scientific method of approach to governmental problems." [100]

For scientific research, at least, a vigorous spokesman in high places had emerged. In an address before the AAAS at the end of 1926, Secretary Hoover noted that the United States was spending about $200 million yearly for applied research and development engaging over 30,000 men. For pure science, however, we spend less than $10 million, with fewer than

4,000 men involved in it—a dangerous disparity. "The day of the genius in the garret has passed, if it ever existed." "The more one observes," he continued:

> The more clearly does he see that it is in the soil of pure science
> that are found the origins of all our modern industry and
> commerce. In fact, our civilization and our large populations
> are wholly builded [sic] upon our scientific discoveries. It is the
> increased productivity of men which have come from these
> discoveries that has defeated the prophecies of science.[101]

In order to sustain our increasing population and at the same time raise our standard of living, "we must maintain the output of our pure science laboratories." [102] Hoover looked to these areas for support: to the government, which ought to endow more pure research at its scientific bureaus; to business and industry, in their own interests; and to private benevolence. "No greater challenge has been put to the American people since the Great War than that of our scientific men in the demand for greater facilities." [103]

In Hoover's grand scheme of cooperative individualism, the role of government becomes that of coordinator—informing and facilitating the efforts of individuals and groups of individuals and preventing the excesses of the old, laissez-faire capitalism. Science and technology played an important part in this plan. If American industry and agriculture were to meet the needs of a war-torn world, burdened with increasing population, constant economic growth was required; this growth could come only from science-based industries, both new and established, and from science-aided agriculture.

> [T]he world is irrevocably committed [Hoover later writes] to
> an eternal quest of further truth, with certainty of endless and
> ever more rapid change as new knowledge is translated into new
> conveniences and comforts. The social relations of mankind
> have already been altered by these changes beyond the utmost
> imagination of our forefathers. Further and more revolutionary
> changes will be wrought.[104]

The National Research Fund and Support for Scientific Research

When Hale and Carty resurrected Hale's scheme for a privately financed national fund for scientific research, it is not surprising that they turned for support to Hoover. As early as 1918, Hale and his NRC colleagues had suggested that private funds be collected for equipment, released time, and assistance.[105] In 1921 or 1922, and again in January 1924, Hale broached the subject of a national research fund with Hoover himself.[106] In March of that year, Hale moved into high gear; he presented the National Academy of Sciences a proposal, for discussion purposes, concerning the raising of research funds. A committee was formed to consider the question. Carty was particularly enthusiastic:

> I am becoming more and more persuaded that the greatest good
> which can possibly be done the cause of science is to arouse a
> proper sentiment among the public. If they only knew, they
> would furnish money by the million and by the billion.[107]

Carty had cause to be optimistic. The leadership of the NRC was particularly well connected, having access through its advisory committees and through personal contact to major centers of business and finance. Operations centered around key philanthropic foundations; not only did the foundations themselves supply considerable money for NRC-approved projects, but they also served as loci for personal contact between the scientific statesmen of the NRC and American business and financial leaders. For example, the trustees of the Carnegie Corporation, a New York–based philanthropic foundation, had for their president in 1920–21 J. R. Angell (chairman of the NRC in 1919–20) and included Merriam (member of the Academy and chairman of the NRC), Carty (NRC, Academy, and so forth); and lawyer-statesman Root. An important advisor to the Carnegie Corporation was Kellogg, Hoover's friend and the NRC's secretary from 1919 to 1931. Recipients of substantial Carnegie funds were Millikan and the California Institute of Technology. The Carnegie Institution of Washington (of which Hale's Mt. Wilson Observatory was a part and of which Merriam was president after 1921) included on its Board of Trustees Carty, Hoover, Root, Julius Rosenwald (an early donor of funds to the NRC), Welch (Academy and NRC), Flexner (Academy and NRC), and Walcott (Academy and NRC). Millikan was an advisor and research associate. Noyes (Academy and NRC), Millikan, and Hale all received considerable funds. The Rockefeller Foundation, too, was tied in closely with the group. Flexner, Rosenwald, and Kellogg had been trustees of the foundation by 1925, and by 1929 they were joined by Angell and Augustus Trowbridge of the NRC.[108]

The same names keep cropping up. All were influential members of the National Academy of Sciences and of the NRC. Merriam, Millikan, Noyes, Hale, and Dunn were, by 1925, officers or members of the Council of the Academy; Hoover, Jewett, Carty, Welch, Walcott, Angell, and Flexner were important members of academy sections. When one turns to the roster of leaders of the NRC, one sees, again, the same names. Chairmen of the NRC included Hale, Carty, Merriam, Angell, and Dunn. Its secretary was Kellogg, and leaders of various divisions included Millikan, Flexner, Trowbridge, Jewett, Hale, Walcott, and Merriam.

It is no exaggeration, therefore, to claim that there existed an NRC establishment that formed an interlocking directorate with the parent National Academy of Sciences and with major foundations. The donors, administrators, and recipients of scientific research funds comprised a closely knit network. Thus, when the National Research Endowment was launched, there was good reason for optimism concerning its rapid success.

There was, of course, much preliminary work to be done. The Academy authorized the NRC to undertake a "movement" for securing research funds. The movement needed a leader. Hale suggested Hoover early on, and both Carty and Hale agreed that "a man of his type is required if anything is to be accomplished." [109]

In view of Hale's courting of Hoover as early as March 1921, his letter to Dunn in November 1925 remains puzzling: "You know of my doubts about Mr. Hoover [Hale wrote] and my complete unwillingness to accept him as chairman unless I could be convinced of his attitude towards fundamental research." [110] Hale's doubts were apparently quickly dissipated, and on November 8, 1925, when the committee reported to the council on the National Research Endowment, Hoover was named chairman. At the Academy's November 9 meeting, trustees were named to direct the Endowment; they included Hoover, Root, Rosenwald, Mellon, Charles Evans Hughes, Colonel Edward M. House, Millikan, Hale, Carty, Dunn, and Welch.[111] The *Declaration of the Trustees* recorded their faith that "human progress depends in large degree upon research in pure science" and proposed "immediately to secure adequate funds for the encouragement of research in pure science."[112]

Hoover threw himself into the Endowment's drive with great enthusiasm. His kick-off speech was given on December 1, 1925, in New York, under the title "The Vital Need for Greater Financial Support to Pure Science Research." The themes were familiar but were now expressed with great vigor. "Not only is our Nation [Hoover claimed] today greatly deficient in the number of men and equipment for [fundamental research] but the sudden growth of industrial laboratories has in itself endangered pure science research by drafting the personnel of pure science into their ranks. . . . Applied science will dry up unless we maintain the sources of pure science."[113] He concluded:

> We have prided ourselves on our practicality as a Nation.
> Would it not be a practical thing to do to give adequate support
> to pure science? And if by chance we develop a little
> contribution to abstract learning and knowledge, our Nation
> will be immensely greater for it.[114]

The New York *Times* featured the Hoover speech and ran editorials strongly supporting the endowment during the early years of the campaign.[115] Newspapers all over the country, from the Tonopah (Nevada) *Bonanza* to the Peekskill (New York) *Star*, printed the pure science publicity sponsored by the NRC's trustees.[116] Initial monetary successes seemed formidable too: Mellon contributed $10,000 for the organizational expenses of the Endowment, and in February 1926, W. S. Gifford of American Telephone and Telegraph pledged $200,000 per annum for ten years, provided the total collected was $20 million—and all that after visits by Hale and Millikan.[117] But Hoover reported to Hale that "We have made some progress but it is evident that we are faced with a long campaign and that this campaign will need more systematic organization than we had thought at the outset." [118]

The Great Engineer set out to provide just that systematic approach. He appointed Carty his vice-chairman and in May set up a committee of information, with regional representatives, for the Endowment. He had long lists of speakers drawn up and laid plans for a radio promotional campaign. His regional representatives included Noyes, J. S. Ames, Jewett, and Kellogg. His list of speakers included such luminaries as G. N. Lewis, Whitney, Langmuir, and Raymond Pearl of Johns Hopkins. Hoover wrote Kellogg that it is imperative to demonstrate the necessity of pure research for progressive industry and that "it is vital to every stockholder in every corporation." [119] Lists of large corporations and suggestions concerning the proper Endowment representatives to approach them were prepared. Dunn, Carty, and Jewett were suggested for major automotive and electrical manufacturers; Welch, Flexner, and Pearl were to help out with pharmaceuticals and insurance companies; E. V. McCollum of Johns Hopkins, who had discovered several vitamins and was an authority on nutrition, was to be asked to contact the baking industry; and Stratton was to contact George Eastman of Eastman Kodak. [120]

By 1927, the Endowment had evolved into the National Research Fund—a more fitting title, since monies were to be solicited on an annual basis for a decade rather than in a single sum. Root had given as legal opinion the view that corporations, while they could not give money to charity, could "invest" in pure scientific research: "Industries created by utilizing the results of scientific research for the benefit of mankind," he wrote, "must continue along the same path or they will fall behind and be superseded by others more farsighted." [121]

Optimism was still in order. "We are all in high spirits," Carty informed Hoover in April 1927. Pledges were made, most of them conditional upon the Fund's receiving the entire $20 million. J. P. Morgan of United States Steel vowed to contribute $100,000 per year for ten years. [122] Eastman pledged $100,000 for five years, if the entire sum could be collected by 1930. "Millikan is a wonder and so is Eastman," Carty cheered. [123] Other donors followed: Julius Rosenwald, the National Electric Light Association, the American Iron and Steel Institute—all pledged large sums, subject to the usual conditions.

By mid-1929, the Fund had received pledges of $900,000 per year for five years. [124] The scientific committee of the Fund had produced a list of worthy research recipients, and among those to receive "tentative allotments" were A. A. Michelson, Karl Compton, P. W. Bridgman, R. W. Wood, James B. Conant, G. N. Lewis, Norbert Wiener, J. J. Abel, and H. S. Jennings. [125] But the structure was already faltering. Hoover left the chairmanship in 1928 to run for the Presidency; his successor, Carty, had to resign at the end of 1929 because of ill health. [126] Carty was succeeded by Jewett, who, as it turned out, remained only to bury the body. With the default of the National Electric Light Association and of Rosenwald, too, the Fund was in serious trouble by 1930. Millikan indicated his desire "to wipe the slate clean of our previous commitments." [127] After the

National Electric Light Association agreed in 1931 to subscribe to only 15 percent of its commitment, Jewett wrote that "the situation of the Fund is indeed a desperate, disheartening, and to me, a rather disgusting one . . . Private subscription is clearly out of the question because of the state of the times." [128] Moribund, the Fund limped on. In 1934, the Academy finally refunded the money (about 94 cents on the dollar) to subscribers of the National Research Fund; the Fund was officially dead.[129]

The Fund withered for complex reasons. Hale, Millikan, and the others badly miscalculated the willingness of American business to invest in an enterprise so far removed from the market. Many corporations in a competitive situation were reluctant to underwrite a free ride for those unwilling to contribute.[130] Helen Wright is doubtless correct in her view that corporate shortsightedness and the Depression finally put an end to the Fund;[131] A. Hunter Dupree is certainly on the mark in asserting that the loss of Hoover and Carty in quick succession did considerable harm.[132] R. C. Tobey has made an important point in charging the leaders of the Fund with an ideological failure; that is, they could not convince many corporate leaders of their vital indebtedness to pure research.[133] But it would be a mistake to believe that the Fund had no lasting impact. The movement to employ private contributions to support science was far from depleted and awaited only more favorable economic conditions for a revival.

Only three years after the death of the National Research Fund, the National Academy of Sciences created a committee to investigate possible philanthropic support for science. The committee included such luminaries of the old National Research Fund as Flexner, Dunn, Jewett, and Millikan, plus Compton and Harvard astronomer Harlow Shapley. In January 1938, two veterans of the Fund, Dunn and Jewett, met with Albert Blakeslee, the botanist and AAAS president in 1940; by the end of the month, their recommendations were forwarded to the Academy. Blakeslee made a special point of investigating and drawing upon the experiences of the National Research Fund before launching the Science Fund.[134] The Science Fund was officially launched in April 1941, under the auspices of the Academy. The executive committee included Blakeslee, Shapley, and Jewett (president of the Academy), and the board of directors listed other important veterans of the old Fund: Angell, Millikan, and Hoover.[135] The coming of World War II in 1941 arrested any growth of the Science Fund, but its activities as coordinator of research activities continued. The scale of the Science Fund, as compared with its predecessor, was modest; before 1941, the largest gift, from the Carnegie Corporation, was $25,000. In 1945, the Sugar Research Foundation gave $50,000 to be used for a prize. Research grants from 1939 to 1948 totaled only $4,400.[136]

Shapley became chairman of the Board of Directors in 1945 and remained chairman in the postwar years of struggle to regain momentum and to create the government-financed National Science Foundation. "An organization completely free of governmental management [Shapley wrote in 1950] is advisable to parallel the National Science Foundation. A nongovernmental agency would not only be free from political pressures but would be more acceptable as a recipient of industrial grants." [137]

The National Science Foundation, created in 1950, spelled an end to the need for the Science Fund. The New Deal, World War II, and what was then generally termed the Atomic Age had tremendously altered the relations between science and society. Jewett could still go before Congress and testify against the National Science Foundation, arguing that tax laws should be liberalized to encourage philanthropy toward scientific research, but his audience was no longer so receptive;[138] the optimism of 1925–26 had suffered at the hands of major events. The New Era had long passed into history and so had its marriage of science and what Hoover had called American individualism.

A Note on Sources

Sources include Herbert Hoover Papers, Hoover Presidential Library, West Branch, Iowa; George Ellery Hale Papers and R. A. Millikan Papers, California Institute of Technology; William Welch Papers and Isaiah Bowman Papers, The Johns Hopkins University; and National Academy of Sciences Archives, Washington, D.C.

Footnotes

1. R. A. Millikan, "The New Opportunity in Science," *Science* **50** (1919), p. 285.
2. R. M. Yerkes, Ed., *The New World of Science: Its Development during the War* (New York: Century, 1920), pp. 39–48.
3. Millikan, "New Opportunity," pp. 289–290; R. A. Millikan, "Some Scientific Aspects of the Meteorological Work of the United States Army," in Yerkes, *New World*, pp. 49–62.
4. C. J. West, "The Chemical Warfare Service," in Yerkes, *New World*, p. 153.
5. Millikan, "New Opportunity," p. 291.
6. *Ibid.*, p. 292.
7. *Ibid.*
8. See, for example, C. Rosenberg, "Science and American Social Thought," in D. van Tassel and M. Hall, Eds., *Science and Society in the United States* (Homewood, Ill.: Dorsey Press, 1966), pp. 135–162; E. Layton, *The Revolt of the Engineers* (Cleveland: Press of Case Western Reserve University, 1971), chap. 3; R. Wiebe, *The Search for Order* (New York: Hill and Wang, 1968), chap. 6.
9. See J. K. Galbraith, *The New Industrial State* (Boston: Houghton Mifflin, 1967), chap. 6.
10. Millikan, "New Opportunity," p. 293.
11. *Ibid.*
12. *Ibid.*, p. 294.
13. *Ibid.*, p. 296.
14. *Ibid.*, p. 297.
15. On Hale, see, for example, D. Kevles, "George Ellery Hale, the First World War and the Advancement of Science in America," *Isis* **59** (1968), p. 437.
16. G. E. Hale, "National Academics and the Progress of Research II: The First Half–Century of the National Academy of Sciences," *Science* **39** (1914), p. 199.
17. G. E. Hale to W. H. Welch, July 13, 1915, The Johns Hopkins University, Welch papers, box 35.
18. W. H. Welch to G. E. Hale, July 14, 1915, Welch papers, box 35.
19. L. N. Scott, *The Naval Consulting Board of the United States* (Washington, D. C.: Government Printing Office, 1920), pp. 10–11.
20. G. E. Hale to W. H. Welch, July 13, 1915, Welch papers, box 35.
21. See G. E. Hale, "The National Research Council," in Yerkes, *New World*, pp. 13–30; Kevles, "Hale," pp. 430–432.
22. G. E. Hale, "National Academics," pp. 680–682.
23. C. K. Mees, "Organization of Industrial Research Laboratories," *Science* **43** (1916), p. 766.

24. On the wartime NRC, see A. H. Dupree, *Science in the Federal Government* (New York: Belknap Press, 1964), chap. 16.

25. Kevles, "Hale," pp. 432–444; Dupree, *Science in Government*, pp. 327–329. For the NRC's attempt to centralize U.S. science see R. C. Tobey, *The American Ideology of National Science since 1919–1930* (Pittsburgh: University of Pittsburgh Press, 1971), pp. 49–61.

26. National Research Council, *Third Annual Report* (Washington, D. C.: Government Printing Office, 1919), p. 26; Kevles, "Hale," p. 435.

27. "Proposal for the Endowment of Research in Physics and Chemistry," California Institute of Technology, Hale Papers, box 452, folder 401. See also Tobey, *American Ideology*, pp. 55–57.

28. See Kevles, "Hale," pp. 435–437; Dupree, *Science in Government*, pp. 229–230.

29. G. E. Hale to J. C. Merriam, September 19, 1922, Hale Papers, box 281.

30. Both letters are quoted in N. Reingold, "National Aspirations and Local Purposes," *Trans. Kansas Acad. Sci.* 71 (1968), pp. 241–242.

31. A good description of the fate of these bills can be found in D. Kevles, "Federal Legislation for Engineering Experimental Stations: the Episode of World War I," *Technol. Cult.* 12 (1971), pp. 182–189.

32. Draft Minutes of the Organizing Meeting of the National Research Council, National Academy of Sciences Archives, Administration: Executive Board: Meetings. Hereafter cited as NAS Archives, A: EB.

33. Report of the Joint Conference of National Associations of State Universities, Land Grant Colleges and Committee of the NRC Regarding the Newlands Bill, November 18, 1916, p. 68, NAS Archives, A: EB: Projects.

34. W. R. Whitney to R. A. Millikan *et al.*, March 16, 1917, NAS Archives, A: EB: Projects.

35. National Research Council Executive Board to P. V. Stephens, April 9, 1917, NAS Archives, A: EB: Projects.

36. P. G. Nutting to P. V. Stephens, December 4, 1918, NAS Archives, A: EB: Projects.

37. G. E. Hale to G. Dunn, December 24, 1918, NAS Archives, A: EB: Projects.

38. G. E. Hale to A. Swazey, October 25, 1917, NAS Archives, A: EB: Projects.

39. Engineering Council, Conference on Smith-Howard Bill, January 7, 1919, p. 1, NAS Archives, A: EB: Projects.

40. A. O. Leuschner to G. E. Hale, February 8, 1919, NAS Archives, A: EB: Projects.

41. R. A. Millikan, "Research in America after the War," *A.I.E.E. Trans.* 37 (1918), p. 1732.

42. A. T. Poffenberger, Ed., *James McKeen Cattell: Man of Science*, vol. 2, *Addresses and Formal Papers* (Lancaster, Pa.: Science Press, 1947), p. 370.

43. R. A. Millikan, "Address of Acceptance of the Norman Bridge Laboratory of Physics," *Science* 55 (1922), p. 331.

44. *Ibid.*, p. 332.

45. R. A. Millikan, "Science and Society," *ibid.* 58 (1923), p. 297.

46. *Ibid.*

47. J. M. Cattell to A. A. Noyes, July 2, 1916, Hale Papers, box 9.

48. R. A. Millikan, "The Practical Value of Pure Science," *Science* 59 (1924), p. 10.

49. R. A. Millikan, "The Relations of Science to Industry," *ibid.* 69 (1929), p. 29.

50. National Research Council, *Third Report*, p. 26.

51. F. Rhodes, *John J. Carty* (New York: privately printed, 1932), p. 65.

52. National Research Council, *Third Report*, p. 26.

53. National Research Council, *Fourth Annual Report* (Washington, D. C.: Government Printing Office, 1920), p. 18.

54. See W. D. Lewis, "Industrial Research and Development," in C. Pursell and M. Kranzberg, Eds., *Technology in Western Civilization* (New York: Oxford University Press, 1967), vol. 2, chap. 40; D. Kevles, *The Study of Physics in America* (unpublished dissertation, Princeton University, 1964), chap. 7.

55. Kevles, *Study*, p. 252.

56. R. A. Millikan, *Autobiography* (New York: Prentice-Hall, 1950), p. 117.

57. See U.S. National Resources Committee, *Research—A National Resource*, vol. 2, *Industrial Research* (Washington, D. C.: Government Printing Office, 1941), pp. 34–35; National Research Council, *Third Report*, p. 26.

58. F. B. Jewett, *Industrial Research*, National Research Council Reprint and Circular Series, No. 4, pp. 2–3.

59. F. B. Jewett, "Industrial Research with Some Notes Concerning Its Scope in the Bell Telephone System," *A.I.E.E. Trans.* **36** *(1917)*, pp. 842–843.

60. O. E. Buckley, "Frank Baldwin Jewett," *NAS Biog. Mem.* **27** (1952), pp. 248–249.

61. Lewis, "Industrial Research," p. 627.

62. National Resources Committee, *Research*, pp. 52–53.

63. Lewis, "Industrial Research," p. 627.

64. New York *Times*, December 27, 1925, sect. 2, p. 6.

65. J. J. Carty, "The Relation of Pure Science to Industrial Research," *Smithsonian Institution Annual Report—1916* (Washington, D. C.: Government Printing Office, 1917), p. 525.

66. *Ibid.*, p. 527.

67. *Ibid.*, p. 531.

68. C. P. Steinmetz, "Scientific Research in Relation to the Industries," *Frank Inst.* **182** (1916), p. 713.

69. *Ibid.*

70. *Ibid.*

71. See, for example, L. Graham, "The Development of Science Policy in the Soviet Union," unpublished manuscript.

72. C. E. Skinner, "Industrial Research and Its Relation to University and Governmental Research," *A.I.E.E. Trans.* **36** (1917), p. 873.

73. *Ibid.*, p. 846.

74. J. J. Carty, *Science and the Industries*, National Research Council Circular No. 8 (1920), pp. 13–14. See also J. J. Carty, *Science and Progress in the Industries*, National Research Council Circular No. 89 (1929).

75. J. W. Williamson, "Industrial Research Laboratories," *Nature* **121** (1928), p. 409.

76. National Resources Committee, *Research*, p. 176.

77. *Ibid.*, pp. 101–103.

78. *Ibid.*, p. 50.

79. New York *Times*, July 16, 1924, p. 16.

80. A good account of governmental research after World War I can be found in Dupree, *Science in Government*, chap. 17; C. Pursell, "Science and Government Agencies" in *Science and Society*, chap. 8.

81. National Research Council, *Fourth Report*, pp. 13–14.

82. National Research Council, *Organization and Members 1920–1921* (Washington, D. C.: National Academy of Sciences, 1922), pp. 11–12.

83. Pursell, "Government Agencies," p. 233. See also R. Cochrane, *Measures for Progress* (Washington, D. C.: Government Printing Office, 1966) chap. 2.

84. Cochrane, *Measures*, pp. 225–226.

85. E. B. Rosa, "The Scientific Work of the Government," *Sci. Mon.* **11** (1920), pp. 249–252.

86. "Elimination of Waste in Industry," *Mining Metal.* **2**, No. 172 (1921), p. 7.

87. H. Hoover, "The Only Way Out," *ibid.* **1**, No. 159 (1920), p. 7.

88. H. Hoover, "Reorganization of the Work of the Federal Government," *Science* **53** (1921), p. 399.

89. G. E. Hale to H. Hoover, March 9, 1921, Hoover Presidential Library, West Branch, Iowa, Hoover Papers, Commerce Official File (Comm OF), box 144.

90. National Research Council, *Organization and Members 1925–26* (Washington, D. C.: National Academy of Sciences, 1925), p. 7.

91. G. E. Hale to H. Hoover, January 15, 1924, Hoover Papers, Comm OF, box 144.

92. F. B. Jewett to H. Hoover, December 12, 1922, Hoover Papers, Comm OF, box 8.

93. G. Dunn to H. Hoover, February 24, 1923, Hoover Papers, Comm OF, box 85.

94. H. Hoover, *Memoirs* (New York: Macmillan, 1952), vol. 2, pp. 83–84.

95. *Ibid.*, p. 73; V. Kellogg, "Herbert Hoover and Science," *Science* **73** (1931), p. 198. Compare, however, Cochrane, *Measures*, pp. 231–233.

96. "The National Bureau of Standards Under Hoover," Hoover Papers, Comm OF, box 85.

97. Department of Commerce, *Fourteenth Annual Report of the Secretary of Commerce* (Washington, D. C.: 1926), p. 22.

98. G. E. Hale to H. Hoover, June 14, 1928, Hale Papers, box 12; G. E. Hale to Allen, undated draft, 1928, Hale Papers, box 12.

99. G. E. Hale to H. Hoover, November 7, 1928, Hale Papers, box 12.

100. R. A. Millikan, speech for H. Hoover, November 7, 1932, California Institute of Technology, Millikan Papers, box 48.

101. H. Hoover, "The Nation and Science," *Science* **65** (1927), p. 27.

102. *Ibid.*, p. 28.

103. *Ibid.*

104. H. Hoover, "The Scientific Work of the Government of the United States," *Sci. Mon.* **36** (1933), p. 7.

105. *Proposal for the Endowment of Research in Physics and Chemistry*, 1918, Hale Papers, box 52.

106. G. E. Hale to H. Hoover, January 15, 1924, Hoover Papers, Comm OF, box 144.

107. J. J. Carty to G. E. Hale, May 29, 1924, Hale Papers, box 9.

108. See Carnegie Institution of Washington, *Yearbook 30* (Washington, D. C.: Carnegie Institution, 1931), p. v; Rockefeller Foundation, *Annual Report* (New York: Rockefeller Foundation, 1929), p. 11.

109. G. E. Hale to J. J. Carty, May 22, 1925, Hale Papers, box 9; G. E. Hale to J. J. Carty, May 21, 1925, *ibid.*

110. G. E. Hale to G. Dunn, November 4, 1925, *ibid.*, box 50.

111. National Academy of Sciences, *Report (1925–1926)* (Washington, D. C.: Government Printing Office, 1927), pp. 16–17.

112. *Declaration of the Trustees of the National Research Endowment*, Hoover Papers, Comm PF, Box 56. See also *Science* **63** (1926), pp. 158–159.

113. H. Hoover, *Vital Need*, p. 4, Hoover Papers.

114. *Ibid.*, p. 11.

115. New York *Times*, December 2, 1925, p. 12; *ibid.*, December 10, 1926, p. 18; *ibid.*, December 29, 1926, p. 2.

116. There is a list of newspapers in the Hale Papers, box 12.

117. Hoover Papers, Comm PF, box 56.

118. H. Hoover to G. E. Hale, February 8, 1926, Hale Papers, box 12.

119. H. Hoover to V. Kellogg, February 27, 1926, Hoover Papers, Comm PF, box 56.
120. Hoover Papers, Comm PF, box 56A.
121. E. Root to H. Hoover, Hoover Papers, Comm OF, box 254.
122. J. P. Morgan to H. Hoover, January 3, 1928, Welch Papers, box 51.
123. J. J. Carty to G. E. Hale, June 18, 1929, Hale Papers, box 9.
124. Tobey, *American Ideology*, p. 215.
125. Minutes of Meeting, National Research Fund, June 18, 1928, Hoover Papers, Comm PF, box 56.
126. J. J. Carty to R. A. Millikan, October 23, 1929, Hale Papers, box 9.
127. R. A. Millikan to F. Jewett, August 12, 1930, Hale Papers, box 50.
128. F. Jewett to G. E. Hale, November 18, 1931, Hale Papers, box 50.
129. F. Jewett to G. E. Hale, April 20, 1934, Hale Papers, box 50.
130. L. E. Davis and D. Kevles, "The National Research Fund: A Case Study in the Theory of Economic Institutions and Industrial Financing of Academic Science," unpublished manuscript, pp. 17–21. Dr. Kevles has kindly made available to me a draft of the paper.
131. H. Wright, *Explorer of the Universe* (New York: Dutton, 1966), p. 369.
132. Dupree, *Science in Government*, pp. 342–343.
133. Tobey, *American Ideology*, pp. 222–223.
134. P. Brockett to A. Blakeslee, February 24, 1939, The Johns Hopkins University, Bowman Papers.
135. *Philanthropy in Science*, Bowman Papers.
136. *The National Science Fund*, Bowman Papers.
137. H. Shapley, "Report to Academy Council on the National Science Fund," NAS Archives, A: EB: Projects.
138. J. L. Penick, *et al.*, Eds., *The Politics of American Science* (Chicago: Rand-McNally, 1965), pp. 84–86.

In his lively recollection of the 1920's, *Only Yesterday*, F. L. Allen notes that "The prestige of science was colossal. The man in the street and the woman in the kitchen . . . were ready to believe that science could accomplish almost anything." Electricity, far more mysterious to the average person than was steam or diesel power to his parents, was the new magic; a scientist investigating the vanishingly small could be received as a beneficent wizard. "The word science," Allen continues, "had become a shibboleth. To preface a statement with 'Science teaches us' was enough to silence argument." The addresses of the presidents of the AAAS in the 1920's began to reflect the scientific community's awareness of its popularity. J. Playfair McMurrich's "A Retrospect" of 1923 (edited here) stressed the progress of the AAAS and reminded his audience of their responsibility to bring the message of the march of science to the man in the street. "The distrust of seventy years ago," McMurrich rather confidently proclaimed, "has given way to trust and the world accepts with tranquillity the shattering of many old beliefs, providing that the necessity for their destruction is vouched for by competent scientific opinion." He saw the new status of science as a reward for the cornucopia that science and technology had provided; the next step, of course, was an appreciation of truth for its own sake.

The scientific method, after all, was democratic. Charles D. Walcott's 1924 address exhorted "each individual to train and conquer himself." Indeed, ideals, aspirations, and consciences of men were subject to science. A man "would be a better business man, a better citizen and more successful . . . if he had a working knowledge of scientific method."

During the 1920's, one of the newest sciences, psychology, had captured the public imagination. Public debate and popular interest in the psychoanalytic techniques of Freud, Adler, and Jung, as well as in Watson's behaviorism, ran high. The psychologist's word, from industrial management to child-rearing, held new authority. J. McKeen Cattell's address of 1925 (shortened here) exhorted the scientists in his audience to claim forthrightly their place in the sun: "Scientific men should take the place that is theirs as masters of the modern world." In fact, the experience of the postwar years proved that "The advancement of science should be the chief concern of a nation that would conserve and increase the welfare of its people."

Yet the rapid pace of technological change took its toll. By the last half of the decade, there was widespread concern that science and technology had outstripped man's abilities to deal with them effectively. When the Bishop of Ripon in 1927 asked for a ten-year 'science holiday' in order to give society a chance to assimilate science's rapid growth, the alarmed reaction was swift and surprisingly vigorous. The scientific community and its allies, among them the New York *Times* (September 7, 1927), strongly denounced the holiday seekers. In the January 1930 issue of *Scribner's Magazine*, R. A. Millikan expostulated on the "Alleged Sins of Science." Frank Jewett spoke of a "senseless fear of science." But the ritual invocation of science had lost much of its magic by 1930, and in the May 1930 issue of *Scribner's* Christian Gauss, dean of Princeton University, could talk of "The Threat of Science" and by indirection charge Millikan with pretentiousness and arrogance.

After the crash of 1929, congressional appropriations for scientific work in the government began to reflect the heightened mood of scepticism toward science, as well as the prevailing mood of fiscal retrenchment. A survey made by *Science Service* in midsummer 1932 showed that pending bills had cut funds for science by about $9.5 million (approximately 12.5 percent), with the Geological Survey (30.6 percent), the National Bureau of Standards (25.6 percent), and the Bureau of Fisheries (32 percent) particularly hard hit. In December, President Hoover submitted his budget estimates, which introduced further cuts, but which left the Bureau of Standards untouched and actually increased appropriations for the Geological Survey.

At the same time, scientists, like many others, found it difficult to obtain work. The sharp rise in research employment between 1927 and 1931 was reversed; a steep decline continued until 1933. Between 1931 and 1933, 110 companies discontinued research laboratories, and over 5,000 laboratory workers lost their posts. In New York, for example, a Committee on Unemployment and Relief for Chemists and Chemical Engineers was established and found laboratory space and materials for the unemployed to carry on research. It was estimated that 20 percent of the chemists and chemical engineers in the New York metropolitan area had been laid off by December 1932, and the numbers were still rising.

Franklin D. Roosevelt's administration brought no immediate respite. Further cuts were made in science and education budgets as the New Deal commenced. Some scientific bureaus faced cuts of over 50 percent of their 1931–32 appropriations. The Bureau of Standards reduced its staff, once at 974, by 380 persons. The protests of leaders of the scientific community and of the New York *Times* (July 21, 1933) against the reductions mattered little.

Within the administration, a new spokesman for science emerged in Secretary of Agriculture Henry A. Wallace. *Science* magazine reported in the June 23, 1933, issue that Rexford Tugwell and Wallace were "making a strenuous fight" for agricultural research funds. *Science* of August 3, 1934, also published Wallace's speech "Research and Adjustment March Together." In it, Wallace underscored a new role for science within his own vision of the New Deal:

> We might just as well command the sun to stand still as to say
> that science should take a holiday. Science has turned scarcity
> into plenty. Merely because it has served us well is no reason
> why we should charge science with the responsibility for our
> failure to apportion production to need and distribute the
> fruits of plenty equitably. Science has done the first job and
> done it magnificently.

The Depression, according to Wallace, had its roots in maldistribution, not in technological unemployment, as antiscience spokesmen had charged. In a tribute to the scientific spirit, Wallace appealed to scientists and engineers to apply the precision of their thinking to social life. He concluded (*Science*, January 5, 1934):

> [W]e wish a wider and better controlled use of engineering
> and science to the end that man may have a higher percentage
> of his energy left over to enjoy the things which are
> non-material and non-economic, and I would . . . particularly
> include the idle curiosity of the scientist himself. Even the most
> enthusiastic engineers and scientists should be heartily desirous
> of bending their talents to serve these higher human ends.

The chairman of the NRC, Isaiah Bowman (later to be president of Johns Hopkins), moved to strengthen the hand of the scientific community, getting in touch with important cabinet members such as Wallace, Harold Ickes, and D. C. Roper. Bowman even suggested that Wallace be elected to the National Academy of Sciences. It was Wallace, too, who passed on to President Roosevelt Bowman's idea for a Science Advisory Board, which was created by Executive Order on July 31, 1933, but which turned out to be short-lived.

The chairman of the Science Advisory Board, Bowman's own preference, was Karl Compton, president of MIT. Compton had for some time been concerned with problems of science policy. In 1927, he swam against the prevailing current in advocating federal support for research by means of a tax on corporate profits, although he conceded at the time that probably "this can never be done by legal means." During the New Deal, those means appeared on the horizon. As head of the Science Advisory Board, Compton became a leading spokesman for science planning and for federal support for research. "Federal policy," he wrote in July 1934, "has been more like that of a fly-by-night promotion enterprise than of a stable enduring business, building for its future." Later that year, Compton reiterated the theme "Should not the public . . . contribute at least a large part of the necessary financial backing?" Interestingly, Compton was as sensitive as Millikan and G. E. Hale to the dangers of federal sponsorship—to wit, the politicization of research—and advocated self-administration of such funds through the Academy and the NRC.

Compton continued to lobby for scientific research. At a meeting of the American Institute of Physics in New York in February 1934, Compton, Millikan, and Jewett spoke out against a tendency to impose an effective moratorium on research through the National Recovery Administration codes. "Science makes more jobs," Compton insisted, not fewer. Compton's Recovery Program of Scientific Progress, a $16 million renewal project, was, however, rejected. The Science Advisory Board itself was allowed to run its course and expired at the end of July 1935. By November 1938, Compton's disillusionment with the New Deal was nearly complete; he deprecated it as playing with the "same old greasy cards."

In his address at the Indianapolis meeting in 1937, E. G. Conklin attempted to meet the objections of the increasingly vocal antiscience critics and to encourage his fellow scientists to concern themselves with the social affairs of men and the role of science within them. While agreeing with the view that there does indeed exist a disparity between social and scientific progress, Conklin sought the answer in the integration of science into what he called "ethical education." He reaffirmed the nobility of the scientific tradition against what he regarded as ill-considered attacks.

Wesley Mitchell's 1939 address likewise saw science as on the defensive. He cited as widespread the tendency to hold science responsible for depression, totalitarianism, and the horrors of war. "For their part," Mitchell countered, "scientific men are appalled at the hideous uses to which their discoveries are put." It is a human frailty, it seems, to shift one's attitude toward science as one personally benefits or suffers from its applications. It is time, however, Mitchell argued, to transcend that parochial view and build some real understanding of the social function of science. The scientific community bears great responsibility as citizens and as investigators—not as "high priests," but as "brave opponents of prejudice and hysteria." Within two years, the coming of World War II drastically altered the social and political context of the scientific enterprise, but Mitchell's address seemed to possess lasting relevance.

A Retrospect

J. Playfair McMurrich
Cincinnati, 1923

James Playfair McMurrich (1859–1939) *was a Canadian biologist whose major research was in animal morphology. After preliminary preparation at the University of Toronto, McMurrich traveled to Baltimore to study at The Johns Hopkins University, where he took his Ph.D. in 1885. He taught for many years in the United States before returning to the University of Toronto in 1907. His textbook* The Development of the Human Body *(first edition 1902) was widely used and went through seven editions. Elected to the presidency of the AAAS in 1922, McMurrich was an active member of many U.S., Canadian, and European societies throughout his long career.*

It is the custom of our association that the annual presidential address should be delivered, not by the actual president, who assumed the responsibilities and honors of his office at the close of the last meeting a year ago, but by his predecessor, who by courtesy is termed the "retiring president" although as a matter of fact he is not "retiring" but "retired." He has to the best of his ability sustained the responsibilities of the presidency and has been relieved of them, he has enjoyed the honor of the position and has retired beyond the range of the spot-light only to be dragged into it once more with even greater responsibilities than before. Professor Dana in his presidential address of nearly seventy years ago describes this situation more eloquently than I can. "In most offices," he said, "the duties terminate with the office, and the thing of the past, the ex-officer, is to the present an unknown quantity. But it is not so with your president. Science . . . sternly drags forward its reluctant presidents to their hardest trial when they have ceased to be, to a judgment after death severer than that of Rhadamanthus." And Professor Asa Gray nearly twenty years later, naturally and happily employing a botanical metaphor, compared the president to a biennial plant: "He flourishes for the year in which he comes into existence and performs his appropriate functions as a presiding officer. When the second year comes round, he is expected to blossom out in an address and disappear."

This arrangement has its advantages in that it affords what should be ample time for the preparation of such an address as the occasion and the position demands; for a speaker from this rostrum is confronted with the responsibility of speaking as one having authority, as a representative of science, and while he may not have the ability to duly mix "reason with pleasure and wisdom with wit," he may be expected to set forth with surety and clarity the faith that is in him as to the achievements and progress of science, or at least of that department of science which he cultivates. Few can see this responsibility approach with cool, calm composure and assurance, and alas! the very fact that one has apparently ample time for the preparation of one's pronouncements with most of us merely leads to the postponement of that preparation until in the fulness of time it is forced upon one. I confess that this has been my own case and that I am one of those "who time gallops withal." I make this statement not in extenuation but in explanation.

But while I fully appreciate the burden of responsibility that rests upon my shoulders it is with a peculiar satisfaction that I appear here tonight. That I have recently held the highest office in the gift of this association is in itself a source of the greatest pride and satisfaction, but these feelings are enhanced by the fact that I am a witness for the broad spirit of catholicity shown by the association in that it declines to recognize geographical boundaries to scientific endeavor. My presence on this platform is the outcome of a recent meeting of the association on Canadian soil and is to be regarded as a compliment to the association's hosts on that occasion, the University of Toronto and the Royal Canadian Institute. We esteem it an honor that we should have been permitted to act as your hosts, we rejoice in such invasions across our boundary, invasions that tend to maintain and strengthen that *entente cordiale* which, with some slight and temporary perturbations, has characterized the relations of the two countries for well over a century. . . .

I call attention to these facts only to emphasize the broad spirit of fellowship that characterizes this association. Its object is the advancement of science, and it is ready to extend the privilege of its meetings and the stimulus that they bring, wherever, upon this continent, they may be welcome. Canadian scientists and Canadian science have always been as welcome at the association's meetings as that brand of scientist and that brand of science that is produced in the United States. Furthermore, arrangements are now on foot whereby it is hoped that the influence of the association in promoting the advancement of science will be extended to the republic that lies south of the Rio Grande and the association is thus justifying its title of American in a fuller and broader sense than that usually attached to that designation. It is working toward the realization of the ideal expressed in its first by-law, which lays down the principle that "The association is American, its field covering North, Central and South America. Inhabitants of any country are eligible to membership." It strives for the advancement of science, wherever cultivated, as a potent factor in civilization.

This broadening out policy is one that has been inherited by the association from its immediate ancestor, The American Association of

Geologists and Naturalists. This association, which was primarily one of geologists, was organized in 1840 as the result of the inauguration of geological surveys of various states of the Union. Those engaged in these surveys felt the necessity for cooperation and discussion that there might be uniformity in the presentation of the results of their work. Soon the zoologists and botanists—the naturalists—were drawn in. Some chemists from the first had been included in the association but, in time, their department became a large and important one and finally the meteorologists found a congenial atmosphere in the association. So the scope of the interests of the association broadened out and at its 1847 meeting, held in Boston, it was decided that it should assume a title more indicative of its scope, and at the meeting of 1848, held in Philadelphia, it became the American Association for the Advancement of Science and in keeping with the new title it extended its membership to include general physicists, mathematicians, economists and engineers.

Thus our association had its beginning more than seventy-five years ago and at the close of its inaugural meeting it had a membership list of 461, an excellent showing, especially when it is recalled that in the early years there were certain restrictions of the membership that were subsequently made less definite. From the beginning, however, it has had the support of the leading scientists of the country; on its first council were Professor Jeffries Wyman, Professor Benjamin Peirce, Professor S. S. Haldeman, Professor Joseph Henry and Professor Louis Agassiz, names that we of to-day recall with reverence and admiration, names that will forever stand in shining letters on the records of scientific achievement in this continent. With such men in control of its affairs success was guaranteed to the young association; it at once became the rallying ground for scientists in all departments of research and in turn attracted those who were interested in scientific progress without taking active share therein. For, as Professor Bache remarked at a later meeting, "Who will say that they do not return wiser, better, more zealous according to knowledge from a meeting—with Henry, Peirce or Agassiz?"

The first meeting is of interest too from the standard of papers presented. Foremost among these was an exhibition by Lieutenant Maury of charts of the North Atlantic showing the prevailing winds and currents, deduced from the study of many thousand old log-books, an earnest of what was to develop later into the classical "Physical Geography of the Sea." Lieutenant Maury demonstrated clearly the relation of intensive scientific investigation to practical results, for his charts indicated that the route usually followed by southbound vessels did not allow them to profit by the most favorable winds and by selecting another route, deduced from his observations, and testing it by a number of vessels, it was found that the passage could be made in three quarters the average time taken by vessels following the older recognized route. The introduction of steam navigation in the years that followed detracted greatly from the direct utilitarian importance of Lieutenant Maury's investigations, but he had laid the foundations for our modern science of oceanography and had established principles that, for a time at least, greatly favored commercial intercourse with distant portions of the globe, especially that between this country and the east and that between Great Britain and her Australian colonies. The voyage from Liverpool to Australia in earlier days usually occupied some four months or more, but in 1854 a sailing vessel, following the course advocated by Maury, made the passage in sixty-three days, and that course even in these days of steam, is still largely followed.

Another important paper presented at the first meeting was that on "The Sediment of the Mississippi River" by Dr. M. W. Dickeson and Mr. Andrew Brown, a summary of deductions based on observations extending over eighteen years, and mention should also be made of a series of papers by Agassiz, fresh from his expedition to Lake Superior, on whose shores his practised eye found abundant evidence of glaciation and whose waters yielded to him a rich harvest of fishes which he could compare with the fresh-water fishes of Europe and those of the Spix collection from Brazil that he had already studied.

One may not linger over this first meeting, nor may one pause to note the many interesting contributions presented at later meetings. The activity of the association in these early days was sufficiently great to warrant the holding of two meetings in each of the years 1850 and 1851 and the first of those of 1851, the fifth of the association, was held in this city in the month of May, under the presidency of Professor Alexander Dallas Bache, the distinguished and efficient director of the Coast and Geodetic Survey. The college professor of the fifties was not the migratory bird he has since become, nor were there then the inducements to extensive peregrinations that now exist. The colleges and scientific institutions were ranged along the Atlantic sea-board, the great State Universities, now such important factors in our scientific progress, had not yet arisen, although the University of Michigan had opened her doors in 1841 with a staff of two professors and with eleven students in attendance. All previous meetings of the association and those of the parent society had been held in cities of the Atlantic coast; the May meeting of 1851 was the first held beyond the Alleghenies and in 1851 a journey beyond the Alleghenies was not one to be lightly undertaken, it was an adventure. It may interest you to-night to hear of the expectations of Professor Henry and of the realities he found in attending the first meeting of the association in this city. He confessed that it was the first time that he had been west of the mountains and went on to say that "He expected to see a boundless, magnificent forest world, with scattered dwellings and log-cabin villages and energetic New England-descended inhabitants; he thought to find Cincinnati a thriving frontier town, exhibiting views of neat frame houses with white fronts, 'green doors and brass knockers,' but instead he found himself in a city of palaces, reared as if by magic and rivaling in appearance any city in the eastern states or of Europe." Professor Henry's expectations might have been realized some fifty years earlier; in the meantime Cincinnati had grown to the stature of the Queen city of the West and with her material progress had not failed to make provision for the cultivation of the arts and sciences in such organizations as the Mechanics Institute, the Academy of Natural Sciences, the Mercantile Library Association and the Young Men's Lyceum of Natural History, all of which Professor Henry mentioned with the remark, "These are the pride of Cincinnati—these her noblest works."

The first Cincinnati meeting was in itself a notable event as the first invasion by the association of what was then still regarded as the West. But it was made still more notable by two other happenings. At the preceding meeting at New Haven, Professor O. M. Mitchell, to whose enthusiasm the erection of the original Cincinnati Observatory was due, and who was its director until 1859, reported that he had invented and constructed two instruments by which in a single night as many accurate determinations of right ascensions or declinations might be made as were made at the Royal Observatory at Greenwich in a whole year. This was rather a startling claim to be made by one working apart and with few of the resources available at the more richly endowed observatories of the East and of Europe, and a committee was appointed with Professor Peirce as its chairman to investigate the claim and report upon it at the Cincinnati meeting. The committee found that as to the apparatus for observing right ascensions the claim was fully justified and while a sufficient number of observations had not been made with the apparatus for determining declinations to warrant a definite statement regarding it, yet it was regarded as being perfectly correct in the principles of its construction. "The committee," I quote from its report, "are not aware that the history of Astronomical Science exhibits a more astounding instance of great results produced by what would seem to be wholly inadequate means. With the ordinary tools of a common mechanic and with insignificant pecuniary outlay an isolated individual has aspired to rival the highest efforts of the most richly endowed institutions—and his aspirations have been crowned with success." The fame of the Cincinnati Observatory was at once established, for the genius of its director had developed methods of observation that were later adopted by all the leading observatories of the world....

Thirty years were to elapse before the association again met in Cincinnati, that is to say, it was not until 1881 that a second meeting was held in this city, this time under the presidency of Professor G. J. Brush, of Yale University. We have seen that the first meeting was made memorable by an important change in the administration of the association; the second meeting was made memorable by the adoption of a new constitution involving some important changes in organization. Up to 1875 two sections had been tacitly if not actually recognized in the association, Section A including mathematics, physics and chemistry, and Section B including natural history. In the year mentioned the disruptive tendencies of specialization began to manifest themselves and the chemists segregated in what was officially termed a permanent subsection, a similar action was taken by the anthropologists, one year later the microscopists decided that their highly magnified world required a subsection for itself and five years later still the entomologists deemed it necessary that they should betake themselves to a special hive. For each of these four permanent subsections there was a chairman, while the presiding officers of the two original sections were designated by the more dignified term of vice-presidents.

By the new constitution adopted at the second Cincinnati meeting the permanent subsections were abolished and at the same time the association was divided into nine sections, each of which was presided over by a vice-president, who was required each year to give an address before his section. The nine sections were those of A, Mathematics and Astronomy; B, Physics; C, Chemistry; D, Mechanical Science; E, Geology and Geography; F, Biology; G, Histology and Microscopy; H, Anthropology, and I, Economic Science and Statistics. Mark the significance of this step. It was a recognition of the tendency toward specialization that had become so marked a feature in the science of the day, and established a policy that has prevailed up to the present. That is why the second Cincinnati meeting was a notable one. We are now entering upon the third Cincinnati meeting—that it too may be a notable one is what may be expected from the past, but whether its notability will depend on new developments of policy or on its records of scientific achievement we must wait to see.

The policy of specialization thus inaugurated in 1881 was bound to lead to further developments. The first modification of it, however, was in a retrograde direction, consisting of the absorption of Section G, Histology and Microscopy, into Section F, Zoology, in 1885. Looking back from our present standpoint it is difficult to understand why this section G was ever established and its absorption was a step to the good. But it was not long until Section G was reestablished by the division of the section of biology into sections of zoology and botany (1892). Then followed in 1900 the establishment of Section K for physiology and experimental medicine and in 1906 Section L was created for education and at the same time the title of Section H was changed to anthropology and psychology. Section M for agriculture was established in 1912 and in 1921 astronomy was divorced from mathematics, psychology from anthropology and new sections for historical and philological sciences and for manufactures and commerce were created, bringing the total number of sections up to sixteen. There are still some letters of the alphabet available for future sections.

Nor was this recognition of specialization the only sign of segregation in the association. A geographical segregation was bound to come as the sphere of influence of the association grew. It has come; for in 1914 a Pacific Division was established and in 1920 a Southwestern Division, each with its own constitution and officers, each holding its own annual meeting and yet remaining bound to the parent association by the closest ties of membership and purpose. If the vision that our first by-law calls up is to be realized, it is evident that other divisions must be recognized, indeed, as has already been indicated, the establishment of a Mexican Division has already come to be a matter for deliberation. And what a vision it is that our first by-law calls up—a federation of divisions extending from the shores of the Arctic Ocean to Cape Horn, marching under one banner and with one purpose, the advancement of science and civilization!

So with segregation integration was also taking place. But the principle of segregation that the association felt itself obliged to recognize was not adopted as extensively as some groups of scientists desired, and a tendency developed for these groups of specialists to form their own societies independent of the association. It became evident that if such secessions went on the representative character of the association would be endangered. Specialization had come to stay; indeed, it was bound to increase with the growth of the very object to which the association was pledged, the advancement of science. There were advantages for these societies in holding their meetings in conjunction with the association and it was to the advantage of the association that they should do so. That the mutual advantages might be ensured to some extent, the council of the association was empowered to enter into relations with certain of these societies, which became designated as affiliated societies and, in time, were granted the privilege of electing one, or in some cases two, representatives to the council of the association. The number of the affiliated societies has grown prodigiously in recent years and amounts to something over fifty, a fact that may be taken as evidence of the success of this line of policy. The strength of the association does not depend alone upon the size of its membership, but this may be taken as an index of the extent to which it is fulfilling its purposes. Beginning with 461, the membership remained in the neighborhood of 500 until 1870, when a marked growth took place, bringing it up to 2,000 in 1885. Then followed a period of rest lasting until 1900, after which a steady and phenomenal growth occurred until now our membership is approximately 12,000. Surely in such figures we may find reason for congratulation and evidence of the wisdom of the policy laid down by the Council and ably carried out by our late permanent secretary, Dr. L. O. Howard and his successor, Dr. Burton E. Livingston.

Specialization must necessarily accompany progress. When one embarks upon a career of investigation one chooses a stream whose prospect pleases and for a time one floats placidly upon its bosom, following up its course. But soon it is joined by a large tributary and one must decide whether one will follow the right or the left branch. The decision made, one continues one's course, passing tributary after tributary, all of which, like the stream that is being followed, lead into unknown lands and at each a fresh decision must be made. In time the current strengthens, the journey becomes more arduous, difficulties are encountered, but still one keeps on, reaching farther and farther into the unknown and farther and farther from fellow searchers who have chosen other branches. One can not join them if one would, for they are ever advancing, perhaps with even greater rapidity and so one must perforce devote himself to the territory before him, hearing only by chance and at intervals rumors of the discoveries that are being made in other areas. That, it seems to me, is the experience of the investigator expressed in metaphor. The farther he and his associates advance, the more they become isolated. New ideas demand new terms in which they may be discussed, and so the members of each group come in time to speak a peculiar language and their isolation thus becomes more pronounced, for there is a limit to the number of languages that each of us can understand; some of us, indeed, have but a moderate command of even our native tongue.

And if this be a true statement of conditions, if it be true that even those familiar with the scientific methods find difficulty in appreciating the work of those laboring in other fields, how much more difficult must it be for those who from choice or from lack of opportunity have not had the advantage of a scientific training and yet are deeply interested in the progress and achievements of science. These form a not inconsiderable and important portion of our membership; they come to our meetings to hear something of the latest achievements of science and they listen to addresses largely in an unknown tongue. They ask for bread and are given a stone and profit little from such a monolithic repast. Yet these are the persons that we should endeavor to interest if we are truly and fully pledged to promote the advancement of science. Esoteric science may lead from discovery to discovery, but until the significance of its discoveries is made intelligible to what are termed the men in the street it fails to secure popular support. The unintelligible is mysterious and mystery awakens either ridicule or dread.

Much has been spoken and written concerning the need for a popularization of science and something has been done towards its accomplishment, notably the establishment of *Science Service* so ably edited by Dr. Slosson. But is not this very thing a prime duty of this association, devoted as it is to the advancement of science, and does the association live up to the full measure of its responsibilities in this matter? I believe I am right in stating that we have not been so successful in this respect as some of the sister associations in other lands. True, we make some endeavor in providing special evening lectures that are designedly popular, and I venture to suppose that the presidential addresses are expected to partake largely of that character. Nor will I be revealing any secrets of policy when I say that the council has given the matter serious consideration, and one may hope that its conjoint wisdom and experience will devise additional means to meet the difficulty. In the meantime it may seem temerarious to suggest measures looking to the betterment of the situation, but a retiring president has privileges and I feel so strongly the necessity for retaining and increasing the interest of what may be termed the lay members of the association in the aims and results of scientific research that I will venture a suggestion. Lack of understanding leads to misunderstanding, and I would beg that those who contribute papers to the sections, and especially the vice-presidents of sections, should in their deliverances bear in mind our lay members and strive for simplicity and perspicuity. Most of us are educators and we have in the meetings of this association opportunities for educating found nowhere else. Let us remember this and take advantage of our opportunities.

These ideas were suggested by the perusal of a number of addresses given by early presidents of the association. There runs through several of them an almost apologetic note, as if it seemed necessary to defend researches into the mysterious phenomena of the universe, since conclusions based on these researches tended to unsettle men's minds by undermining old long-standing beliefs. This was three generations ago and the practical applications of science were neither so frequent nor so striking as they are to-day. The Morse telegraph had been used commercially four years before the first meeting of the association, but the other remarkable applications of electrical energy that have become so much a part of our every day life were as yet undeveloped. Anesthesia had been introduced into surgical practice, but antisepsis, that was to revolutionize surgery, was as yet unknown; indeed, the causation of sepsis, together with that of putrefaction and fermentation, was awaiting an explanation by the genius of Pasteur, and this explanation was to lead up not only to surgical antisepsis, but to the formulation of the germ theory of disease and the wonderful achievements of modern preventive medicine. How these and other achievements in other departments of science have revolutionized the world! They are tangible evidences of the benefits that science can confer upon mankind, they are recognized as such by the man in the street and he consequently has developed an interest in science and a toleration of its votaries that his forebears of three generations did not possess. Nay, not only does he tolerate science, he encourages it by providing funds for its prosecution, by richly endowing great research laboratories and by bequeathing princely prizes as rewards for important discoveries. The distrust of seventy years ago has given way to trust, and the world accepts with tranquillity the shattering of many old beliefs, providing that the necessity for their destruction is vouched for by competent scientific opinion. The theory of relativity, whether or not its full significance is understood, is swallowed without a spasm, even though it may displace the theory of gravitation from what seemed to be its unassailable position; and that the atom, supposed to be the ultimate, indivisible abstraction of human thought, is in reality a more or less complex system of electrons revolving planet-like about a central nucleus, even this idea is accepted without a tremor.

This change of attitude is undoubtedly largely due to an increased appreciation of the value of science as shown by its practical applications. This may not have been the only factor, but it is a potent one. It is impossible to consider the multitudinous and marvelous facilities that have become parts of our daily life without realizing that they are but the practical applications of scientific principles to the control or utilization of natural forces and materials, without, in other words, perceiving that it is to scientific investigation that we are indebted for these advantages. The men who have made these practical applications become known and respected, their names become household words, they are the representatives and high-priests of science and their glory is reflected upon even the most abstruse fields of scientific investigation. The attitude assumed may be expressed thus: "See what great benefits science has conferred! It promises others and therefore it is to be encouraged."

For the present we must perhaps be satisfied with this. For several centuries science was under the ban, dogma was supreme and science, which necessarily found itself in contest with this, was impious and heretical. Truth was standardized and complete, and to question that accepted truth was to undermine the foundations of belief. The human mind is conservative in its reactions; habits of thought are as difficult of modification as habits of action, and the change from the dogmatic to the scientific habit has been slow; indeed, it is far from complete even now. The utilitarian appeal of science has done much to emancipate it from its thraldom to dogma, but it is not yet universally recognized that the utility of science depends absolutely upon its success in discovering truth. It is only by getting at the tiue facts and the true principles involved in any problem that the results of science become useful. The scientist is a searcher after truth and it is for that reason that he is able to confer benefits on humanity; it is for that reason that he deserves recognition. Surely he should feel no necessity for an apology for his existence.

But the ultimate truth is elusive. When science establishes a truth that may seem at first to be ultimate it but points the way to another truth lying beyond, and it is to the credit of scientific men that they are ready to admit the lack of finality in what has been accomplished, once the vista of the new truth has opened out. This attitude is not easily understood by the layman unfamiliar with the scientific method, and he is apt to imagine that a confession of lack of finality means the condemnation of the older truth as false. This is a misconception that has frequently occurred and, unfortunately, it is a misconception that scientists themselves have aided in creating, by failing to appreciate the popular view-point. In the popular mind the doctrine of evolution is so completely involved in Darwin's exposition of it that it has come to be regarded as the product of his brain. Consequently any acknowledgment that some of Darwin's views may require modification is assumed to imply that the foundations of evolution are shaken. It seems trite to repeat once more the true relation of Darwin's theory to the doctrine of evolution, but there seems to be need for its repetition. Evolution as a theory long antedates Darwin's time; Laplace, to go on farther back, found it in the history of the heavenly bodies. Lyell demonstrated it in the history of the earth, and Goethe, Saint Hilaire and Lamarck saw it in the history of terrestrial organisms. What Darwin did was to give a plausible and convincing explanation of how organic evolution might have occurred, but whether that explanation is or is not the correct one matters not so far as the doctrine of evolution is concerned; that stands unshaken even though Darwin's explanation of how it was brought about be discarded. The evidence in its favor to-day is many times stronger than it was in Darwin's time, and it seems incredible that man as a reasoning animal should presume to doubt its validity; such doubts can be based only on ignorance of the evidence or on unreasoning prejudice.

True, it was Darwin who focussed the attention of the world upon the doctrine, by propounding the theory of natural selection as the causal factor in the transmutation of species. The biological world of to-day does not ascribe to that factor the importance that Darwin gave it. Its action can not be denied; it is self-evident to any observer of nature's way who finds

> that of fifty seeds
> She often brings but one to bear.

It plays an important rôle in the suppression of the unfit rather than in the survival of the fittest, but it can act only on variations sufficiently pronounced to determine life or death. It has been shown in several cases that what seem trivial variations may, under certain conditions, lead to fatal results, but even admitting these, it is difficult to believe that many of the minute differences that distinguish species have selective value. Natural selection acts effectively in the perpetuation of species, but it does not originate them and to that extent the modern biologist may depart from Darwin's standpoint. Darwin was looking for the origin of species; the modern biologist goes a step further and is looking for the origin of variations and the mechanism of heredity problems far beyond Darwin's times. But he stands on the foundation built by Darwin, since the whole structure of modern philosophy rests on that foundation. . . .

With new methods the fields of investigation have broadened out and knowledge has increased by leaps and bounds. And it is with especial satisfaction that we may note that in these progressive zoological studies the scientists of this continent have always been well in the van, if not in the forefront of the advancing column. But all through the most overwhelming flood of new knowledge there runs the guiding clue supplied by the doctrine of evolution. That has been the stimulus and dominating idea in all these studies; without it many, very many of them would never have been conceived and knowledge would have lost thereby. No! Evolution is not dead, nor can it be killed by legislative enactment.

Let me conclude this retrospect with a message for guidance in the future, taken from one who did not always find satisfaction in the advances and applications of science, and all the more impressive on that account. It is not the first instance in which a prophet from whom curses might be expected gave blessings, real or implied, instead. The words are those of Mr. Ruskin. "Go to Nature," he says, "in all singleness of heart and walk with her laboriously and trustingly, having no other thought but how best to penetrate her meaning; and remember her instructions—rejecting nothing, selecting nothing and scorning nothing; believing all things to be right and good and rejoicing always in the truth."

Science and Service

Charles D. Walcott
Washington, D.C., 1924

Charles Doolittle Walcott (1850–1927) *was a world-renowned geologist and science administrator. Virtually self-educated beyond the secondary level, he had been fascinated by tribolite fossils and rocks as a youngster; five years of extensive field research on a farm in central New York launched his professional career in 1876 as an assistant to American paleontologist James Hall. When the U.S. Geological Survey was established in 1879, Walcott joined the staff, where he remained for 28 years, the last 13 years as director. His geological and paleontological research centered on Cambrian rocks and fauna throughout the United States and Canada.*

Surprisingly, Walcott's field work did not diminish appreciably as administrative responsibilities multiplied. Director of the Geological Survey after John Wesley Powell, head of the National Museum, and secretary of the Smithsonian Institution from 1907 until his death, he was instrumental in founding and organizing the Carnegie Institution (1902), the U.S. Reclamation Service (1903), the Research Corporation (1912), the National Advisory Committee for Aeronautics, forerunner of NASA (1915), and the National Research Council (1916). Walcott was an eminent figure in the National Academy of Sciences and recipient of numerous honors from universities and scientific societies in the United States and abroad. He served as president of AAAS in 1923.

The service of science to humanity was initiated when the early philosophers began to discover and record the more simple truths of nature. Then, as now, the "sciosophists," defined by David Starr Jordan as "apostles of systematized ignorance," rejected all evidence without attempting to prove or disprove, and condemned the exceptional mind that was compelled by an inner urge to attempt to penetrate into the mists that concealed the unknown.

Through many centuries human thought was confined largely to channels predetermined by the religious and political concepts of the more intelligent. To this privileged group freedom of expression by others was intolerable, and a seeker after truth other than that provided and established by authority was fair game to be hunted down and destroyed. Progress came as an undercurrent of truth that slowly gathered strength and volume, until, like the warm waters of tropical ocean currents penetrating the colder regions, it weakened the congealed crust of ignorance and prejudice, and, combining with the sunshine of scientific research, gradually dispersed systematic opposition to the growing appreciation of natural phenomena and laws.

The story of the onward march of scientific research in quest of truth is not unlike that of the white race in the conquest of America. A few fearless souls penetrated the wilderness and blazed the trails that others might follow; some fell martyrs to their zeal, but more came after; facts were determined, laws established, which in time contributed to the welfare of the race.

Science has made great progress, but its conquest of the unknown and of the unprogressive doctrines of the sciosophists is still far from complete. When we consider the many vital problems that are being discussed and the indifference and limited training of a large proportion of supposedly intelligent people, it seems only well under way. The search for truth and the interpretation of facts and their verification and application must continue, with ever-enlarging conceptions, until the ultimate destiny of our race is fulfilled.

Millikan has well said: "The purpose of science is to develop without prejudice or preconception of any kind a knowledge of the facts, the laws and the processes of Nature." The search for truth based on this scientific method is one of the greatest inspirations and incentives to high and thorough endeavor. The researcher may be engaged in any field of human knowledge, including religion, history, literature, natural science or government; his motive may be of the highest type of pure delight in investigation and discovery for the sake of truth, or it may be sordid and egotistic; but his results, after being subjected to the exacting tests of general discussion and criticism, will find their place in the general scheme of the advancement of the physical, mental and spiritual well-being of man.

The scientist need not enter into controversies with the theologian or the sciosophist. He may make mistakes and interpret nature incorrectly, and he will never become so infallible or omniscient as to be sure that he has the entire or exact truth, but, as in the past, he may add a little here and a little there, until a great inspired mind grasps the multitude of facts and interprets from them a general fundamental law of nature. A structure of theory or hypothesis is then reared that may endure for a generation or two, until a continually growing knowledge requires some modification that may strengthen it, or lead to its partial or complete rejection.

An illustration of work to be done by the specialist may be drawn from my own investigations relating to primitive life. In 1876 I spent a month tracing the distribution of the "Potsdam Sandstone" about the Adirondack Mountains of New York, and in collecting fossils from the sandstone and the immediate superjacent Hoyt limestones. I was told the following winter by the geological authorities of the day that there was little use of attempting to discover traces of life in rocks lower than the Potsdam, as they rested upon the Archean. During the forty-eight years that have passed since then I have always had in view the discovery of older and still older evidences of primitive life. The Upper, Middle and Lower Cambrian faunas were all found in orderly succession beneath the Potsdam, and finally bacteria and algal deposits far below in the pre-Cambrian. Sixteen years ago I discovered a locality in the lower Middle Cambrian Burgess shale of the Canadian Rockies in which the fossils were marvelously preserved. This deposit yielded a wonderful series of marine invertebrates showing clearly that at this early time the evolution of the life in the sea had progressed far beyond our previous conceptions of it, and that, with the exception of the cephalopods and vertebrates, there has been very little development in the classes of animal life during the millions of years between the time of the Burgess shale fauna and the present. Recently there have come to light fossils in the upper portion of the Lower Cambrian that indicate that the types of the Burgess shale fauna were then in existence.

Beneath the Lower Cambrian on all the continents there is a great unconformity between the basal beds and the superjacent pre-Cambrian formations that have been grouped in the Proterozoic and Archeozoic eras. Although we have found traces of bacteria and algal growth in the Proterozoic, it is not considered that these forms were the progenitors of the Cambrian faunas. In fact, the search for primitive pre-Cambrian evidences of life other than that which may have existed in fresh water lakes and streams has come to a standstill at the great unconformities beneath the Cambrian and the Proterozoic. I do not know of a trace of the great primal marine life of the ocean basins from which the Cambrian faunas are descendent, except the few fragments that have been found in the Proterozoic; nor have we found evidence of the deposits of undoubted marine sediments on the present land areas that are older than those of the lower Cambrian. This leaves the field open for the research student of the future to discover the earlier forms of life that in a slow evolution through millions of years finally developed into the highly organized invertebrates of early Cambrian time.

Research students in all branches of scientific endeavor soon discover that with every accession of knowledge new problems open up that will demand attention and effort for years to come. I recall full well as a young man consulting with older scientists and getting the impression that they thought a large proportion of the great scientific problems had been settled; and they often denounced most bitterly the younger men who dared to question the accepted conclusions of their scientific studies.

The sciosophist will rail at and denounce the scientific method, as he can not or will not comprehend it, but it is clearly the only method by which the errors of the present and truth of the future may with certainty become known. "Beyond the conclusion reached through competent study of nature and natural law by cautious and highly trained investigators, it is not possible for any one to go, except as the searcher after the truth himself pushes further ahead in his quest."

Such a broad organization as the American Association for the Advancement of Science must consider ways and means by which it may do its part in the great readjustments of thought and action constantly going on, which will aid or obstruct the advance of mankind towards a more just and ever finer civilization. In view of the accumulation of scientific data and the number of active research workers in all advanced countries, what can our association and its members do to be of the greatest service? As stated by our constitution, the objects of the association are "to promote intercourse among those who are cultivating science in different parts of America, to cooperate with other scientific societies and institutions, to give a stronger and more general impulse and more systematic direction to scientific research, and to procure for the labors of scientific men increased facilities and a wider usefulness." Our membership already includes trained and amateur scientists, educators and intelligent laymen, about 13,000 of them, and it is increasing with the years more rapidly in proportion than is the population of the United States.

For three fourths of a century the members of the association have contributed individually and in groups to the advancement of many branches of science, and some of them have more or less successfully undertaken to interpret to the average intelligent person the story of nature so far as it has been made known by scientific research. All these activities have been helpful, and valuable service has been rendered by the association as a whole, especially by bringing together large groups for the consideration and discussion of the scientific problems of the day.

The title of the association implies association for a definite purpose, "the advancement of science." This in our modern life means also the physical, mental and moral advancement of the human race, not only our immediate associates and fellow Americans, but every human being on the earth's surface, and, for that matter, every form of life that is not predatory or inimical to the welfare of humanity. Is the association now doing its part in the formation of public opinion and in the guidance of social movements; is it in touch with the layman, or is it a professional organization that is slipping away from its great opportunity of informing and arousing the interest and enthusiasm of the American people and the intelligent people of all countries? These questions are important, for to-day a story of science clearly and simply told is welcomed by the press, the author of textbooks, the lecturer, the encyclopedist and agencies for the dissemination of information in all nations.

The association should act as a liaison agency between professional science and the public. It also should act as the liaison agency between the various sciences. The greater the specialization the greater the need of cooperation. In the advance of science, siege tactics have taken the place of guerilla warfare of former days. The big industrial establishments like the Western Electric Company, the General Motors Research Corporation, the General Electric Laboratories and many others show how much more can be accomplished by the combined efforts of experts in various fields of science. This must be carried much farther than an occasional symposium or joint meeting between different departments of the association. The important problems of the day require the cooperation of men from various sciences. For instance, to make an inventory of future energy supplies of the world would require cooperative study by geologists, chemists, physicists, engineers, biologists and sociologists.

More attention must be paid to the unity of nature in order to counteract the artificial divisions which are the natural result of specialization in research. We need men and women who can tell the story of life as a whole, as a zoological story, and who can combine the parts and picture its evolution, character and broad relations, and this applies as well to all scientific, historical and social fields of research.

There is something wrong when ten minutes is the limit for presentation of a valuable paper, with practically no time available for discussion. A few papers, ably discussed, and *leave to print* for others would probably add more to the value and interest of the program than a multitude of papers hurriedly presented, with no opportunity for discussion. For the irreconcilable specialist, the special society or the local general society is open. Again, the poorly prepared paper and even the masterpiece ineffectively delivered is an almost criminal action when we consider the time and effort involved in getting a large group into one place to hear and consider it. Such a loss of time and energy should not be imposed on professional men or laymen. A method should be devised to correct it. Modern life has developed great traffic and social problems that will ultimately be solved, and the association will solve its problems if a group of vigorous, clear-thinking, trained members take them up. I say members, as I think the segregation of women scientists would be an injury to science and the association. Will you attack these problems or will you sit idly by and permit your opportunities to pass into other hands? This was done once and the great work of the National Research Council was initiated and developed by a trio of men who had vision, purpose and inflexible determination to be of service to their country and humanity.

Is the association doing all that it could in connection with the great problem of conservation? Do we realize that our future national progress and the progress of all peoples will be determined, not so much by the acquisition and appropriation of new resources, as by the degree to which we are able to perpetuate and more efficiently to utilize those we already have? Such progress is absolutely dependent upon scientific research and education; it would be difficult to name a single branch of science that is not concerned. The research workers and educators who are members of this association can render no greater service than by concentrating an increasing amount of their effort on the multitudinous phases of this problem.

The United States' unprecedented growth and her present commanding economic position have been made possible by abundance of natural resources. Individual and public economic policies have been predicated on this abundance. Minerals, forests, fur and game animals, agricultural soils, range lands, fish, and water resources were all seemingly inexhaustible in supply, and all have been appropriated and exploited recklessly and wastefully. The cream has been skimmed, and, all too often, the milk has been thrown away.

The whole philosophy of exploitation has been based on the theory of making maximum profits for the exploiter, rather than the ideal of greatest service and lasting benefit to the people of the world as a whole. The resources seemed unlimited, and it was assumed that future requirements would adjust themselves automatically and that posterity would take care of itself.

Now the point has been reached where it is evident that the resources have a limit. Expansion can not continue indefinitely, nor can even the present scale of consumption be maintained as population increases, unless steps are taken to replenish the supply. The pressure of scarcity and increasing costs of exploitation demand the elimination of wastes, the intensive utilization of the resources that are left, and the discovery or creation of new supplies.

A large percentage of our good agricultural soils have been appropriated, and the further expansion of crop production to feed our growing population must come largely through utilization of the poorer land or through more intensive cultivation and fertilization of existing farms. Even more is this true of our pasture and range lands, the *per capita* area of which has been reduced by almost one half since 1890.

Using almost as much timber as all the rest of the world combined, the United States passed the highest point of *per capita* consumption nearly 20 years ago. Even now four times as much is consumed as is grown each year, and only one fifth of the forest land is set apart definitely for timber production. In spite of the growing shortage of timber and the mounting costs of bringing it from remote regions, scores of millions of acres of once productive forest land are lying idle, and we are still wasting two thirds of all the wood that is cut.

The story of our wild life and our waters is little different. Birds, fish, shell-fish, fur bearers, game animals, all have reached an alarming stage of depletion as a result of destructive exploitation. Streams, lakes and coastal waters have been polluted. Many of the streams and lakes which could afford a perpetual source of food, power, irrigation water, recreation, water for drinking, sanitation and other domestic uses, as well as channels for cheap transportation, have been reduced in flow or filled with sediment, following forest destruction or unwise cultivation or pasturing on their watersheds.

All of these are renewable resources. With wise use none of them need have been depleted, and most of them can be made even more productive than they have been in the past. Few would go so far as to contend that such replenishment is unnecessary or undesirable. Many, however, consider it impossible, and even assert that major reductions of the waste in utilizing existing resources are impracticable. The reasons are said to be economic: more intensive farming will not pay, reforestation is too slow and costly, there are no profits in utilizing waste materials.

Yet economic impracticability is frequently only a longer name for ignorance. The discovery of new principles or new methods may make it economically practicable to intensify farming. Better understanding of silvicultural principles and closer study of the life history of our forests will show us how to utilize that resource without jeopardizing its continued productivity, and without increasing the economic burden on the users. Thorough technical knowledge of the product, whether farm crops, timber or what not, will enable us to utilize profitably a great deal that is now wasted.

Our mineral resources, as a general proposition, can not be renewed through human effort, at least in the present state of knowledge. But even with them, the available supplies can be extended almost indefinitely through the discovery of new methods of extraction, or through the discovery and utilization of substitute materials.

To obtain the results desired it is evident that the great masses of humanity have yet to be educated in the scientific method of thought and action, not only in darkest Africa, but here in the United States and in all countries. This is the greatest task immediately before us. All scientific men and women may do their bit—*first*, by training themselves to observe accurately, to think straight and then to record clearly and honestly, and to draw warranted conclusions based on the facts presented, "free from previous preconception and prejudice"; *second*, by reviewing the mass of technical information with which they are familiar and telling the story they have learned in simple, clear language, free from obscure, complicated, technical and verbose wording. These simple suggestions apply not only to research workers in science, but to all the professional classes as well, theologians, doctors, lawyers, statesmen—especially lawyers and politicians, and of course professional teachers in schools and colleges.

That the scientist should have the virtues of charity, tolerance, broad-mindedness, patience, persistence and a very high regard for his fellow man is absolutely essential if he is to reach the heights and be of the greatest service. Agassiz and Pasteur were great scientists and great souls, and gave service by teaching as well as by their example of living on a high plane of thought and action. Some other men have been brilliant contributors to knowledge, although their general manner of living may have been an injury rather than a service to mankind. We need to be grateful for the constructive service of each life, and our criticism of those who have passed and of those who are still active needs to have a broad friendliness as its basis. I believe, too, that a good scientist should be a good Christian, and a good Christian should be a good scientist in his method and work, as both are seeking the truth and the fundamental principles underlying their respective fields of endeavor.

Besides the necessity for each individual to train and conquer himself and to exercise such influence as may be possible on those within his immediate environment, there is great need for him to engage in cooperative public work, by associating with others of similar aspirations, and bringing legitimate influence to bear on all agencies that are concerned in any way with the educational system of the people, from the kindergarten to the university, from the leaflet of the advertising promoter to the great newspapers, magazines and books that make up the thousand and one publications of our day. His influence must also be brought to bear upon the important visual agencies of the motion-picture screen and every other form of illustration, as well as on all those agencies that are seeking to develop "the consciences, the ideals and the aspirations of mankind." The scientific method must be applied to all these factors if we have faith in its ideals.

Is it not practicable for the association to organize a progressive, live committee of men and women to deal with the popularizing of scientific knowledge? It might arrange special sessions for the public to which the layman could go with the feeling that they were for his entertainment and his instruction and not solely to arouse the interest of specialists in their particular field of research. Of all human beings, the child is the greatest and most active investigator of all that pertains to his environment. Why not provide for a junior section of the American Association, and last and in some respects the most important, a woman's section and sessions, at which all the scientific problems of peculiar interest to woman could be considered? We have a strong nucleus of women members, but they should be one of the great influences within the association for developing and carrying forward its work. Then there is the much discussed business man, who has a more or less hazy conception of science and scientific method, depending on whether he considers it affects his interest for good or evil. He would be a better business man, a better citizen and more successful in all his relations in life if he had a working knowledge of scientific method and principles at his command.

Every member of our association should work individually and collectively according to his or her opportunity and ability in supporting the scientific method and in insisting that, in all education of every kind and degree for all classes, the purpose is to develop without prejudice or preconception of any kind a knowledge of the facts, the laws and the processes of nature in all natural and human relations. The natural weakness and incompleteness of all things of human origin will frequently baffle, mislead and confuse, and may even apparently bring temporary defeat, but in the long run there is no other way to eradicate sciosophy, advance the physical, mental and moral welfare of the race and justify our existence and opportunities for service as sentient human beings.

The Pilgrim fathers knew little of science, but they brought the great principles of law, truth, freedom and faith in God to America. Are we doing all in our power to perpetuate and develop them in connection with the multiplex activities of the world of to-day?

Some Psychological Experiments

J. McKeen Cattell
Kansas City, 1925

James McKeen Cattell (1860–1944) *was a distinguished experimental psychologist and energetic science publisher. Cattell's advanced training included work with Francis Galton in England, with G. S. Hall at The Johns Hopkins University, and with the German psychologist Wilhelm Wundt at the University of Leipzig, where he received the Ph.D. in psychology in 1886. Cattell held one of the first chairs in psychology in the United States, at the University of Pennsylvania, until 1891 and developed a unique laboratory research program as head of Columbia University's psychology department for the next 25 years. His own research centered around reaction time studies, psychophysics, and individual differences; although he founded no "school," Cattell's emphasis on the experimental and quantitative approach to psychological research won him professional acclaim as the first psychologist named to the National Academy of Sciences (1901) and the first to serve as president of the AAAS (1924).*

Beginning with the Psychological Review *and* Science *in 1894, Cattell's management and editorial work spanned half a century and included seven major scientific journals, six editions of the biographical directory* American Men of Science, *the formation of Science Press, and trusteeship of Science Service. His influence on the organization of American science was widespread: he was instrumental in bringing together the AAAS and* Science, *participated in the founding of the Psychological Corporation and the American Psychological Association, was actively concerned with problems of university administration and working conditions in academic life, and served on numerous executive committees of the leading scientific societies throughout his life.*

. . . The most important work for psychology and its most useful applications are the measurement of individual, group and racial differences, and the determination of the extent to which these depend on native endowment and on subsequent experience. Indeed it is arguable that this is the most pressing problem of science and of society. If each of us from the moron to the federal president were selected for the work that he can do best, the work fitted in the best way to the individual and the best training given to him, the productivity of the nation would be more than doubled and the happiness of each would be correspondingly increased. If the best children were born, and only they, the welfare of the world would

57

be advanced beyond the reach of practical imagination. Truly the harvest is large, but the psychologists are few. As I said in 1896 in my presidential address before the recently established American Psychological Association: "We not only hold the clay in our hands to mould to honor or dishonor, but we also have the ultimate decision as to what material we shall use. The physicist can turn his pig-iron into steel, and so can we ours; but he can not alter the quantities of gold and iron in his world, whereas we can in ours. Our responsibility is indeed very great." . . .

Formerly a casual acquaintance who learned that he was speaking to a psychologist most frequently inquired about what goes under the name of psychical research—mediums, spirits, ghosts, clairvoyance, telepathy and the rest—often continuing with his own remarkable experiences. Now psychoanalysis is the more usual topic for such after-dinner conversations. Perhaps it need only be said here that witches in New England were convicted on better evidence than can now be adduced for any supernormal events, that miracles at Lourdes are better attested than any of the queer experiences of Sir Oliver Lodge or Sir Conan Doyle. Psychoanalysis is not so much a question of science as a matter of taste, Dr. Freud being an artist who lives in the fairyland of dreams among the ogres of perverted sex. In reference to applied hypnotism, suggestion and psychoanalysis, the remark attributed to Franklin may be repeated: "There is a great deal of difference between a good physician and a poor physician, but not much difference between a good physician and no physician."

Galton in England and Wundt in Germany initiated studies of association, when one word or idea suggests another. In 1880 I published the results of some 15,000 measurements of the time of such associations together with classifications and correlations. These were the first of the psychological measurements of school children which have now been found so useful that some three million tests were made last year in our public schools. The experiments of forty years ago also have a certain interest because American, English, Irish and German students were tested in their native schools, and we thus have a beginning of racial comparisons of which there is urgent need at the present time in view of their practical application to immigration and other social problems of national importance. We have as yet scant scientific knowledge of the differences between the sexes or among races, family stocks, social classes or occupational groups, and of the directions in which, by native endowment or by habits, traditions and institutions, one is different from or superior to another.

In these experiments on association it was found from 363 students in a London school that the time of the total process decreased from 11.76 seconds in the third form, where the children were of the average age of 12.7 years, to 4.13 seconds in the fourth form, where the average age was 17.8 years. There was a positive but small correlation between the tests and class rank, and it was remarked: "It is possible that such experiments measure the alertness of the student's mind more accurately than does the class-rank." This is indeed the fundamentally important aspect of the intelligence tests now used for entrance to college and in many other directions. The psychologist wants to tell what a man can do rather than what he has learned, what use he will make of opportunity rather than what advantages and privileges he has had. The intelligence tests used by psychologists with 1,800,000 conscripts during the war proved most useful in selecting the officer material at the upper end of the curve of distribution, in eliminating those unfit even for the ranks at the lower end. We may look to a great extension of these tests in industry, as soon as they are adjusted more exactly to the conditions, and employers learn their economic value.

In experiments on association we are concerned with the character as well as with the time of the process. The association may be primarily of objects given together in the physical world or it may be due to rehandling by inner experience. Thus it was found that persons engaged in teaching and writing had a larger proportion of logical and verbal associations than others. The Irish students had a larger proportion of these than the English students; the students in a German Gymnasium, where the languages were their main occupation, a still larger proportion of verbal associations. Anecdotes concerning association and the making of categories and classifications have been an occupation of psychologists from Aristotle to the present time. In the paper referred to, statistical methods were used to classify thousands of recorded associations, but it was noted that it was extremely difficult to observe by introspection the process of association, whether in the usual course of mental life or in such experiments. Determinations of free associations are used in psychiatry and in the detection of crime. But in the various types of intelligence tests, it is not the character of the associations, but the time measurements and the correctness of the answers that have proved useful.

Associations and trains of ideas are supposed to be given in terms of the senses. The studies by Fechner and Galton of mental imagery by the questionnaire method were its first use in classifying individuals. Great differences were reported in the clearness and brightness of the images; people have been divided into visuals, audiles, motiles and tactiles. Imagery was one of the original series of tests made on the students of Columbia University in the early nineties, and doctorate dissertations by Dr. Lay and Dr. Betts contain extensive experimental investigations. It appears that people are not competent to describe the train of their ideas and that indirect methods show the comparative unimportance of imagery, which may be only a surface sensory coloring. From tests on Pillsbury, who played twenty games of chess without seeing the boards, it was found that he did not depend on mental images, but on the histories of the games. The situation is different with so-called motor images, for these are really incipient movements and can be measured in the laboratory. All stimuli tend to discharge movements and it seems that one of the principal differences between ideas or images and perceptions is the larger motor element in the latter which gives them what Hume called their "force and liveliness."

There are certainly dreams and hallucinations, and in ordinary perception more of what we seem to perceive is supplied by the central nervous system as molded by past experience than by the incoming currents. Memory images of various kinds and entoptic phenomena form nearly a continuum between imagination and perception. An after-image impressed on the retina or brain twenty years ago is still visible; this may seem incredible, but Newton acquired an after-image that lasted three years. The duration and character of the after-image as well as its color and oscillations were correlated by Dr. Franz with the quality, duration, intensity and area of the stimulus. There are significant individual differences which may denote ability in observation and power of attention.

Ordinary observation, recollection and general information are more defective than is commonly supposed, and this is particularly the case when the person at the time of an event does not know that he will be called upon to describe what has happened. An attorney can discredit a witness by asking questions to which a correct answer should not be expected. As a result of a study published in 1896, it was found that when students in a psychological class were asked—being allowed thirty seconds for a reply—what the weather was a week ago the answers were: clear 16, rain 12, snow 7, stormy 9, cloudy 6 and partly stormy and partly clear 6, about the probable distribution of weather at the beginning of March. When asked which way the seeds in an apple point, 24 said upward, 18 towards center, 13 downward and 3 outward. Answers as to the date of Victor Hugo's death ranged from 1790 to one maintaining that he was still living. Estimates of a given period of 35 seconds ranged from 5 to 150 seconds. When the students were asked what was said during the first two minutes of the lecture in the same course given one week before, the accounts were such that the lecturer might prefer not to have them recorded. From the testimony of the students, it would appear that two minutes sufficed to cover a large range of psychological and other subjects, and to make many statements of an extraordinary character. Similar experiments were made on classes in the Horace Mann School, and these were among the first psychological experiments made on school children.

Practise, learning, memory and fatigue can to great advantage be made the subjects of objective and quantitative psychological experiments. Ebbinghaus's monograph on the learning and recall of non-sense syllables (1885) and the study of the practise curve in learning the telegraphic language by Bryan and Harter (1897) are the classical foundations in these fields. The first practise curve was, however, made in 1886–7, by running daily three miles and plotting the decreasing times and rate of the heart. These also antedated Mosso's experiments on muscular fatigue, on which and its relation to mental conditions many experiments have been made.

How we learn, the best way to learn, the right age at which to learn different things, the transfer of learning from one field to another, are subjects of fundamental importance in psychology and in education. Making practise curves is itself an excellent educational method. The child learns more by working as hard as he can for a short time than by dawdling for a couple of hours. When he plots the curve he is anxious to improve each day's record and the objectively measured competition is with himself as well as with others. Thus it was found that if a boy of ten writes on the typewriter by proper methods twenty minutes a day for sixty days and plots the curves of the amount accomplished and of the errors, he learns to write faster than any one can write by hand. In the meanwhile he learns to spell and to correct his mistakes; he learns arithmetic and geometry as realities; he learns the value of measurement and objective standards. Practise and learning experiments, including records on each of 365 consecutive days, have been made by me for forty years and are now being made. Whatever the scientific value may be, it adds to the interest to keep practise curves in chess, cards, billiards, tennis and the like.

Daily, weekly and seasonal curves; the optimum periods for definite tasks and for a day's work; industrial fatigue; temperature, ventilation and humidity; the most desirable sexual relations, food, amount and distribution of sleep; rest, play and physical exercise; the use and misuse of emotional excitement and of drugs as sedatives and stimuli; these have been the subject of many investigations in the Columbia laboratory. There are none more important in their practical application to the affairs of daily life. The human psycho-physical organism has through long ages by natural selection or otherwise adjusted itself to the world in which it lives. It was not adapted to the innumerable new demands of modern civilization. It has proved itself plastic to an extraordinary degree, but it is the business of psychology to obtain scientific knowledge of the whole situation and then to apply it for the benefit of all.

The Taylor system initiated a new profession of psychological and industrial engineering. It has been retarded because trade unions, not without reason, feared speeding-up methods. A long correspondence with Mr. Gompers shortly before his death indicates that the unions may ultimately in their own interest take up questions of the psychological selection of men and the improvement of the methods of their work. The British Institute of Psychology has been successful in securing the cooperation of the workers and has in some directions increased production by 40 per cent with decreased fatigue. In every field of activity from the use of pick and shovel, of typewriter and ledger, through the factory and office, to the organization of the work of the executive or the congress of the nation, investigations might be made which if put into effect would add from 10 to 100 per cent to effective productivity and lessen to an equal extent effort and fatigue.

It is absurd that researches whose economic value can only be told in billions of dollars and whose contribution to human welfare is even more immeasurable should await the pleasure of a few academic psychologists who take them up in the intervals between coaching the members of a junior social and athletic club and helping with the family housework, and then only until they get into difficulties with the president or themselves become presidents. In our competitive and capitalistic system services to an individual or corporation are paid for, often to excess, whereas services to society are paid for only in the fiat currency of reputation, titles, degrees and the like. A surgeon may receive a thousand or ten thousand dollars for saving or killing his patient. If after years of research he should discover a cure or prevention of appendicitis or cancer, he not only would not be paid for his work, but would lose all future fees. The psychologists of the country, as is becoming for those directly engaged in the study of human behavior, have taken the lead in forming a Psychological Corporation whose objects are to conserve for research part of the profits from the applications of our science and to conduct new research on an economic basis. Scientific men should take the place that is theirs as masters of the modern world.

. . . Psychology, like the other sciences, may date back to Aristotle and as much further as records go; but the earliest research laboratory of psychology was established by Wilhelm Wundt at Leipzig in 1879. The first laboratory courses for students were conducted by me in 1887 and 1888 at the University of Cambridge, the University of Pennsylvania and Bryn Mawr College. When in 1881 and later I worked with Wundt, who was professor of philosophy and lectured over a wide range of philosophical disciplines, he held that the object of psychological experiment was to improve the conditions of introspection and that only the psychologist could be a subject in the laboratory. Early experiments led me to adopt the points of view that psychology, on the one side, is concerned with conduct as well as with consciousness, behavior being the more open to experimental investigations and their useful applications; on the other side, that individual differences are of primary importance, both for constructive science and in the practical affairs of daily life.

There are able psychologists who like to narrate what they think they think, what they feel they feel, what they imagine they imagine. Those of us who are concerned with quantitative measurements and objective results wish them satisfaction, even though we may think, feel or imagine that such literary diversions contribute about as much to a science of psychology as similar stories about their rheumatism and other bodily ailments would contribute to a science of pathology. Some good souls may derive comfort from arguing that they are or have good souls. With Browning's Cleon each might like to say:

> And I have written three books on the soul,
> Proving absurd all written hitherto,
> And putting us to ignorance again.

Philosophy may offer God, freedom and immortality, and it may make some difference whether it does or does not. But the shades of metaphysics wander about the Elysian fields in those obscure regions beyond the river Styx to which the light of science does not reach. Hamlet said: "There are more things in heaven and earth than are dreamt of in our philosophy." There are certainly more things dreamt of in philosophy and in some kinds of psychology than are in the heavens above, or in the earth beneath, or in the waters under the earth; but they do not give us a science of psychology.

In spite of the limited value for science of direct introspection, our mental life is part of the real world, and is that part which is of the greatest concern to each of us. It may be, as has been suggested, that psychology lost its soul long ago and is now losing its mind; but it can not lose consciousness. Our perceptions, thoughts, intentions and feelings are not only elements in sensori-motor arcs; for us they are the end to which the whole creation moves. As far as production goes, consciousness may be only a spectator; but it is the ultimate consumer. We shall have in due time a scientific psychology of human welfare, of the things that are beautiful, good and true. But it will not come until we get these things instead of talking about them; for science has meaning and value only in its usefulness. Psychology may supply economic values equal to those of the physical and biological sciences, human values of even greater significance.

Scientific research and the applications of science in the course of 150 years have increased fourfold the productivity of labor; they have doubled the length of life. Science has made it possible for each to work at routine tasks half as long as formerly and at the same time to consume twice as much wealth as formerly. Fourteen hours of labor, shared by women and children, once provided hovels, lice and black bread for most people, luxury for a few. Seven hours of labor will now supply comfortable homes, warm clothes and healthful food for all. If the resources provided by science were properly distributed—as they will be when we have an applied science of psychology—there is now sufficient wealth to enable all to share in the desirable luxuries that science has created, and to enjoy in full measure the most nearly ultimate goods of life—home, friends, things to do, freedom, self-respect.

The better lives secured through the increased wealth provided by science, together with the applications of science to hygiene, medicine and surgery, have doubled the length of life. In the nations of the west pestilence and famine have lost most of their terror; of the three evil fates only war survives from a prescientific and barbarous past. Much is crude and ugly in the modern world; atrophied instincts and aborted impulses must be replaced by the products of a science of psychology before living can become free and fine. But those who call our industrial civilization materialistic and ignoble have narrow thoughts and scant idealism. They fail to imagine what it means in terms of love and suffering that of ten infants born, formerly only two or three survived childhood, while now eight or nine may live to have children of their own.

The applications of science have abolished slavery and serfdom, the need of child labor, the subjection of woman; they have made possible universal education, democracy and equality of opportunity, and have given us so much of these as we have. Science has not only created our civilization; it has given to it the finest art and the truest faith. The advancement of science should be the chief concern of a nation that would conserve and increase the welfare of its people.

Science and Ethics

Edwin Grant Conklin
Indianapolis, 1937

Edwin Grant Conklin (1863–1952) ranks among the foremost American zoologists of the early 20th century. At Ohio Wesleyan University (B.A., 1886; M.A., 1889) Conklin first formally studied science. With deepening interest in biology, he pursued advanced studies at The Johns Hopkins University, carrying out biological research on cell lineage at the newly founded Marine Biological Laboratory at Woods Hole, Massachusetts. After receiving the Ph.D. from Johns Hopkins in 1891, Conklin taught biology at Ohio Wesleyan, Northwestern University, and the University of Pennsylvania; in 1908 he became department chairman at Princeton University, where he stayed throughout the remainder of his long academic career. Conklin's research fields were embryology, cytology, and the mechanisms of human evolution; many of his experiments, for example, probed the effects of environmental changes on cell structures.

In addition to his work at Princeton, Conklin was instructor and then trustee at Woods Hole for more than 40 years and served many years as trustee and advisor to the Bermuda Biological Station for Research and to the Wistar Institute of Philadelphia. An active member of the American Philosophical Society, Conklin was elected to the National Academy of Sciences, to the presidency of the AAAS (1936), and to offices of specialized scientific societies.

. . . I know full well that there are many scientific specialists who maintain that science has no concern with ethics, its sole function being to seek the truth concerning nature irrespective of how this truth may affect the weal or woe of mankind. They may recognize that the use of science for evil threatens peace and progress, but they feel no responsibility to help avert disaster. The world may be out of joint, but they were never born to set it right; let the shoemaker stick to his last and the scientist to his laboratory.

During the dark days of the world war, I once spoke to a distinguished scientist of some major event in the course of the war and he looked up from his work and said sharply, "What war?" Concentration upon our various specialties is essential, but it should not cause us to lose our sense of orientation in the world. It is pleasant and at times necessary to avoid "the tumult and the shouting," but there is no excuse for the scientist who dwells permanently apart from the affairs of men. At the present time it is probable that nothing else so deeply concerns the welfare and progress of mankind as ethics.

In the early years of the association a favorite theme in the annual address of the retiring president was the relation between science and religion, and pious but more or less futile attempts were made to harmonize "Geology and Genesis" or "Evolution and Revelation." To the majority of modern scientists nothing is more dull and fruitless than such attempts to make science the handmaid of theology, nothing more futile than sectarian conflicts over theological dogmas and creeds and ceremonies. But there is an aspect of religion with which science is vitally concerned, namely ethics, and this has been well called "the religion of science."

Science, as we all know, is tested, verifiable, organized knowledge; ethics is concerned with ideals, conduct and character. Any program looking to human welfare and betterment must include both science and ethics, and there would be great gain for the world if organized religion and organized science could cooperate more effectively in the promotion of practical ethics.

Among the generalizations of science which have been charged with the weakening of ethics, first place must be given to the theory of the natural evolution of man and of ethical systems. It is a fundamental postulate of modern science that man is a part of nature and that his body, mind and social relations have undergone evolution in the long history of the human species. This is not a mere hypothesis but an established fact, if anything is a fact. There is positive evidence that in long past times there were types of human and partly human beings that were much more brutish in body, mind and social relations than the general average of the present race. There is abundant evidence that ethics has undergone evolution no less than intelligence; it has developed from its beginnings in the primitive family group, to tribal, racial, national and international relations; from the ideals and practices of savagery to those of barbarism and civilization; from the iron rule of vengeance and retribution, "an eye for an eye, a tooth for a tooth," to the ideals of love and forgiveness and that highest conception of ethics embodied in the Golden Rule. But as in physical evolution there are retarded or retrogressive individuals and races, so also in the development of ethical ideals some people and periods are far behind others, and all fall short of their highest ideals.

As is well known, the distinctive principle running through the whole of Darwin's philosophy of evolution is what he called natural selection. Having studied the notable effects of human selection in the production of new breeds of domestic animals and cultivated plants, he sought for some comparable process operating in nature without human guidance. This he found in the Malthusian principle of overproduction of populations, the elimination of the less fit and the preservation of favored races in the struggle for life. In general, he regarded the environment, whether organic or inorganic, as the principal eliminator of the unfit, although he assigned a certain rôle to the organism itself as selector and eliminator, especially in sexual selection, while in mental, moral and social evolution this auto-selection played even a larger part in his philosophy.

I shall not at this time discuss the present status of Darwinism further than to say that from practically every branch of modern biology it continues to receive confirmation and extension, so that in spite of severe attacks from many sources and assurances from some excited opponents that "Darwinism is dead," it is still very much alive.

It has been charged by many humanists that Darwinism is destructive of the highest ethical ideals. It is said to be the apotheosis of cruelty and selfishness, to recognize no values except survival, no ideals except success. In this struggle for existence the weak go under, the strong survive; and this is said to justify personal and class strife and wars of conquest. Militarists and dictators have seized upon this principle as justification of their philosophy that might makes right. Conflicts and wars are said to be both the means and measure of progress, and military training to be the highest type of discipline. By both militarists and humanists Darwinism has been considered as an eternal struggle, a vast battle of living things with one another and with their environment, a grim portrayal of

> Nature red in tooth and claw
> With ravine.

Bernard Shaw has said that if Darwinism were true only knaves and fools could bear to live.

This is, however, a fundamental misconception of natural selection. Darwin himself repudiated this extension of his principle to the struggle between races and nations of men. In a letter to Alfred Russell Wallace he wrote that "the struggle between races of men depends entirely on intellectual and moral qualities." Those who attempt to extend the principle of natural selection into the field of intellectual, social and moral qualities should remember that the standards of fitness are wholly different in these fields. Physically the fittest is the most viable and most capable of leaving offspring; intellectually the fittest is the most rational; socially the fittest is the most ethical. To attempt to measure intellectual or social fitness by standards of physical fitness is hopelessly to confuse the whole question, for human evolution has progressed in these three distinct paths. Man owes his unique position in nature to this three-fold evolution, and although the factors of physical, intellectual and social progress are always balanced one against another, they are not mutually exclusive. All three may and do cooperate in such manner that each strengthens the other.

And this leads to the inquiry whether human or so-called artificial selection is not also natural. If we define "natural" as that which is regular and lawful, and not arbitrary and lawless, then human selection is also natural, and this must necessarily follow if man in his entirety is the product of natural evolution. Since Darwin's day the study of the behavior of lower organisms as well as that of human beings in all stages of development from the infant to the adult has shown that selective activity is everywhere present. One-celled plants and animals respond positively to some stimuli, negatively to others, and in general, though not invariably, this selectivity of response is beneficial. For example, they avoid extremes of heat or cold, they move or grow toward certain chemical substances and away from others, they take in as food certain substances and reject others. Even germ cells show some of these same properties, and in general it may be said that all living things manifest differential sensitivity and reactivity, and that by a process of trial and error and finally trial and success they generally manage to eliminate reactions that are not satisfactory and to persist in those that are. This is the Darwinian principle extended to the reactions of organisms in which the organism itself is eliminator and selector. Intelligence in animals and man is arrived at in this same way, by many trials and failures and finally trial and success, remembering of past failures and successes, elimination of the former and persistence in the latter. A cat that by trial and error has learned to open the door of a cage, as in Thorndike's experiments, or a horse that has learned in the same way to lift a latch and open a gate is intelligent with respect to that one situation; intelligence in human beings is acquired in the same way. Indeed, intelligence is the capacity of profiting by experience, while the ability to generalize experiences and to recognize fundamental resemblances in spite of superficial differences is what we call abstract thought or reasoning.

In his famous Romanes address at Oxford in 1892 on "Evolution and Ethics" Professor Huxley maintained that ethics consists in opposing the cosmic process of natural selection by intelligent human selection and in replacing the ruthless destruction of the weak and helpless with human sympathy and cooperation. He illustrated the superiority of human selection by pointing out the fact that a cultivated garden left to nature grows up to weeds and, therefore, that human intelligence can improve on the blind processes of nature in meeting human needs.

All this is undoubtedly true; we are continually improving on nature for our own purposes; all agriculture, industry, medicine, education are improvements on nature. The notion that nature is always perfect is certainly false, and the cry, "Back to nature," is more likely to be a call to regress than to progress. But it is a mistake to suppose that human intelligence and purpose, social sympathy, cooperation and ethics in general are not also parts of nature and the products of natural evolution. In Darwin's theory the environment eliminates the unfit organism, but in individual adaptations to new conditions the organism itself eliminates many useless or injurious responses. In such cases the organism rather than the environment is the eliminator or selector, either by the hit-or-miss process of "trial and error" or by the vastly more rapid and less wasteful method of remembered experience, that is, by intelligence. Thus intelligence can improve on the blind processes of nature, because it is not blind, although it also is natural. And thus intelligence has become a prime factor in evolution. Intelligence and social cooperation have become the most important means of further human progress.

Will and purpose are similarly natural phenomena growing out of the use of intelligence in finding satisfaction. Will is not an uncaused cause but rather the product of all those bodily and mental processes, such as appetites, emotions, memory and intelligence, which stimulate, regulate or inhibit behavior. Ability to thus control activity in response to remembered experience is what we call freedom from fixed, mechanistic action. Both intelligence and freedom vary greatly in different animals and in the same individual at different states of development. They are relatively slight in human infants, but they rise to a maximum in normal adults. However, men are never perfectly intelligent nor absolutely free, but the more intelligent they are the freer they are.

All this is pertinent to a discussion of the natural history of ethics, for social ethics assumes the ability and the responsibility of individuals to regulate behavior in accordance with ideals and codes of conduct. It, therefore, demands freedom to choose between alternatives that are offered. Without such freedom there can be no responsibility, no duty, no ethics. It has long been the creed of certain rigidly mechanistic scientists that freedom, responsibility and duty are mere delusions and that human beings are automata, thinking, feeling and doing only those things which were predetermined by their heredity and environment over which they have no control. This fatalistic creed was in large part a deduction from the determinism of nature which was revealed in mathematics, astronomy, physics and chemistry and was then extended by certain physiologists to all vital phenomena, including human life and personality. Indeed some of these "hard determinists" went so far as to maintain that the whole course of human history was predetermined in the original constitution of the universe, that nations had risen and fallen, cultures and civilizations had come and gone and that the present state of the world and its future destiny were all determined by inexorable laws. However, many biologists who investigated the behavior of animals refused to regard them as mere automata, and students of human behavior generally held that there must be some flaw or break in this logical chain that bound man helpless on the wheel of fate, some fallacy in the logic that denied all freedom and responsibility to man, some monstrous error in the conclusion that saints and sinners, philanthropists and fiends were mere pawns or puppets in a game in which they were moved by forces over which they had no control.

As a way of escape, mathematicians and physicists, who were most impressed by the determinism of inanimate nature, were generally inclined to adopt some form of Cartesian dualism, which would endow living beings and especially man with an immaterial principle or soul which was not subject to this rigid determinism. But on the other hand, students of life phenomena in general could find no sufficient evidence for such dualism, and hence arose the strange anomaly of physiologists and psychologists being more rigid determinists, so far as life and man are concerned, than students of the physical sciences.

Several scientists recently have expressed the view that Heisenberg's principle of indeterminacy in the sub-atomic field can somehow be converted into the novelty, creativity and freedom manifested by living things. But, so far as I am aware, no one has shown how this can be done, since the principle of indeterminacy does not apply to molecules or masses of matter, and living things are always composed of complex aggregations of these. Furthermore, biologists generally do not admit any fundamental indeterminacy in the behavior of living beings. Novelty, creativity and freedom, wherever their origin has been traced, are found to be caused by new combinations of old factors or processes, whether these be atoms, molecules, genes, chromosomes, cells, organs, functions or even sensations, memories and ideas. By such new combinations of genes and chromosomes and environmental stimuli all the novelties of heredity and development arise. There is good evidence that even psychical properties, such as intelligence, will and consciousness, emerge in the process of development because of specific combinations of physical and psychical factors. This is, indeed, the whole philosophy of evolution, namely, that the entire universe, including man and all his faculties and activities, are the results of transformation rather than of new-formation, of emergence rather than of creation *de novo*.

Freedom does not mean uncaused activity; "the will is not a little deity encapsuled in the brain," but instead it is the sum of all those physical and psychical processes, including especially reflexes, conditionings and remembered experiences, which act as stimuli in initiating or directing behavior. The will is not undetermined, uncaused, absolutely free, but is the result of the organization and experience of the organism and in turn it is a factor in determining behavior. Therefore, we do not need to import from sub-atomic physics the uncertain principle of uncertainty in order to explain free will. The fact that man can control to a certain extent his own acts as well as phenomena outside himself requires neither a little daemon in the electron nor a big one in the man.

Just one hundred years ago the English poet, William Wordsworth, wrote:

> Man now presides
> In power where once he trembled in his weakness;
> Science advances with gigantic strides,
> But are we aught enriched in love and meekness?

These lines are much more significant to-day than when they were penned. The strides of science have never been so gigantic as during the past century. So far as our knowledge of and control over natural forces and processes are concerned we live in a new world that could not have been forecast by scientists and could scarcely have been imagined by poets and seers of one hundred years ago. Within the last century we have passed from the "horse and buggy stage" to the locomotive, the automobile and the airplane era; from slow mails to the telegraph and telephone and radio, from education and music and art for the favored few to a time when these are available to untold millions. Applied scientific knowledge has made amazing advances in all the means of living; in the abundance and variety of food and clothing; in comfort, convenience and sanitation in housing; in relative freedom from degrading drudgery and a corresponding increase in leisure and opportunity for the pursuit of happiness. At the same time medical science has to a great extent removed the fear of "the pestilence that walketh in darkness"; no more do whole cities flee in panic from the black death, or yellow fever, or white plague; no more do civilized people live in dread of smallpox or typhoid fever or diphtheria; the average length of life has been greatly increased; physical pain has been reduced and comforts have been multiplied.

These are only a few of the marvelous advances of science, most of them within the memory of old persons still living. No similar progress can be found within any other century of human history. "But are we aught enriched in love and meekness?" With man's increased control over the forces of nature there has not gone increased control over human nature. Man's conquest over outer nature has outrun his conquest over his own spirit, and consequently the gifts of science, which might be unmixed blessings if properly used, become new dangers when used for evil purposes. Science is organized knowledge, and knowledge in itself is neither good nor bad but only true or false. That which gives social and moral value to science is the purpose for which men use it. If it is used for selfish advantage it may weaken or destroy social cooperation. If used for greater and more terrible wars it may end in the destruction of civilization itself.

Neither in human nature nor in social relations has progress kept pace with science. This is not the fault of science but rather of man and of society. The great advances in the applications of science have often been used for selfish purposes rather than for social welfare. Scientific progress in medicine and sanitation is far in advance of its social utilization, but not in advance of its urgent need. Rational and peaceful means of solving class conflicts and of preventing wars would be vastly less costly and more effective than strikes and armaments. Scientific control of population and the necessaries of civilized life would be more humane and progressive than to leave these to the law of the jungle. The fact is that social progress has moved so much slower than science that one might say that scientific progress is matched against social stagnation.

Many thoughtful persons are asking: "Will science, which has so largely made our modern civilization, end in destroying it? Has it not placed powers in the hands of ignorant and selfish men which may wreck the whole progress of the race?"

It is a fact that improvements in human nature are not keeping pace with increasing knowledge of and control over outer nature. By means of language, writing, printing, the radio and all the means of communication and conservation of knowledge each human generation transmits its acquirements to succeeding ones. Thus present-day science, culture and civilization represent the accumulated experience and knowledge of all the past, each succeeding generation standing, as it were, on the shoulders of preceding ones. Every individual, on the other hand, begins life where all his ancestors began, namely, in the valley of the germ cells; he then climbs to the summit of maturity and goes down into the valley of death. But society, gifted with continuous life, passes on with giant strides from mountain top to mountain top. And so it happens that science and civilization in general outrun individual heredity, for the learning and acquirements of each generation are not transmitted to succeeding ones through the germ cells (except in the case of Professor McDougall's trained rats) but only through social contacts. For this reason increasing knowledge and power have greatly outrun improvements in inherent human nature, so that man is still, in the language of Raymond Fosdick, "the old savage in the new civilization."

It is impossible to halt the march of science except by destroying the spirit of intellectual and political freedom. No scientific moratorium by international agreement is possible, even if it were desirable, and any nation that undertook to halt the progress of science would be doomed to the fate of Ethiopia and China. Is there any way of escape from this perilous situation, in which knowledge and power have outrun ethics? Can world-wide ethics keep up with world-wide science? Can science itself do anything to close this widening gap between lagging human nature and the increasing responsibilities of civilization?

Eugenics has been proposed as a possible and necessary solution of this problem. Undoubtedly great improvement in human heredity could be effected, if the principles of good breeding which are used with such notable results in the improvement of domesticated animals and cultivated plants were to be used in the breeding of men. There is no doubt among students of heredity that by means of a system of selective breeding a healthier, longer-lived, more intelligent type could be developed and the prevalence of emotional instability and neuroses could be decreased. But the difficulties in the way of such a eugenical program are enormous where the human stock is so mixed, as it is in almost all races of men, and where the rules of good breeding would have to be self-administered or imposed by authorities that are influenced by social, racial or ethical prejudices. Even if these obstacles could be overcome and this program wisely and persistently followed it would take thousands of years to bring about any marked improvement in the masses of mankind, and in the present crisis of civilization we need a more quick-acting remedy, if it can be found.

Fortunately there are other and more rapidly acting remedies for this disharmony between biological and social progress. Heredity determines only the capacities and potentialities of any organism, the realization of those potentialities depends upon development, which is greatly influenced by environment, hormones, health or disease, use or disuse, conditioned reflexes or habits. In every individual there are many capacities that remain undeveloped because of lack of suitable stimuli to call them forth. Since these inherited potentialities may be social or anti-social, good or bad, it is the aim of enlightened society to develop the former and to suppress the latter. In the heredity of every human being there are many possible personalities; which one of these becomes actual depends upon developmental stimuli. Each of us might have been much better or much worse characters than we are if the conditions of our development had been different. Endocrinologists and students of nutrition are already preventing or overcoming many of the deficiences or defects that arise in the course of development. Medicine and sanitation have notably reduced the occurrence, spread and mortality of epidemics and there is every reason to expect that the causes and cures of the most serious diseases that now afflict mankind will be discovered, that sickness and suffering will be greatly reduced and that the average length of life will be still further increased. In all these respects science is contributing greatly to human welfare and to practical ethics.

But of all the possible means of rapidly improving social conditions, ethical education is probably the most promising. Education, based upon a knowledge of the principles of development and aimed at the cultivation of better relations among all classes, races and nations is the chief hope of social progress. The most enduring effect of education is habit formation. Good education consists in large part in the formation of good habits of body, mind and morals. Heredity is original or first nature; habits are second nature, and for character formation and social value they are almost if not quite as important as heredity itself. Ethical habits especially are dependent on education, and in all normal human beings it is possible to cultivate habits of unselfishness rather than selfishness, of sympathy rather than enmity, of cooperation rather than antagonism. To trust entirely to heredity to improve men or society is to forget that heredity furnishes capacities for evil as well as for good, and to disregard the universal experience of mankind that human nature may be improved by humane nurture.

On these grounds certain humanists have proposed that art, literature, history and political and moral philosophy should replace science in the educational program, since, as they assert, science neglects or destroys the real values of life, inasmuch as it is said to be materialistic, non-ethical and lacking in high ideals. The president of the University of Chicago has recently called science a failure in the educational process and has urged a return to philosophy as the only sure road to sound discipline and true culture. Those who have never experienced the discipline and inspiration of scientific studies fear that science will destroy our civilization, and they call upon educators to repent and to return to the good old subjects of classical learning. Without discussing the specific value of different subjects in the educational program it may be remarked that it was not science that caused the decay of former civilizations nor was it in the power of classical art, literature and philosophy to save those civilizations. The fact is well attested that science has given us grander and more inspiring conceptions of the universe, of the order of nature, of the wonderful progress through past evolution and of the enormous possibilities of future progress than were ever dreamed of in prescientific times. And as an educational discipline there are no other studies that distinguish so sharply truth from error, evidence from opinion, reason from emotion; none that teach a greater reverence for truth nor inspire more laborious and persistent search for it. Great is philosophy, for it is the synthesis of all knowledge, but if it is true philosophy it must be built upon science, which is tested knowledge.

> To the solid ground
> Of nature trusts the mind that builds for aye.

Education, then, which looks to the highest development of the physical, intellectual and moral capacities of men is the chief hope of human progress. Even any possible program of improvement of inherited human nature must rest upon education concerning the principles of heredity and the methods of applying them to the breeding of men. Without waiting for the slow improvement of human nature through eugenics great progress can be made toward the "good society" by the better development of the capacities we already possess. All the advances from savagery to the highest civilization have been made without any corresponding improvement in heredity. Within a few generations, through the inculcation of better social habits or fashions, there have been many improvements in human relations. The torture and execution of heretics, whether theological or political, had all but disappeared from the earth until the recent revival of intolerance under dictatorships; belief in witchcraft and demoniacal possession and methods of exorcising devils by fire or torture no longer exist; human slavery as a legal institution has been abandoned everywhere; in this country the duel is no longer regarded as the necessary way of defending one's honor. These and a hundred other improvements in social relations have come about through education and enlightened public opinion. May we not hope that class, racial and national conflicts and wars may be outmoded in the same way?

Sensations, emotions and instincts are the principal driving forces in our lives as well as in those of animals. Primitive instincts, or what we properly call the "Old Adam," may cause persons, classes and nations to disregard reason and to give way to an orgy of passion. Lawyers for the defense sometimes call this a "brain storm," but it might more truly be called a "brainless or endocrine storm," for it is the sort of behavior which one sees in decerebrate cats or in animals in which the lower centers of the emotions and reflexes are very active but are imperfectly controlled by the higher centers of intelligence and reason. One of Europe's dictators says, "We think with our blood," which is a pretty sure way "to see red." Another dubious test of truth is "to feel it in the bones," which is generally indicative of ossified thought. It is especially man's superior brain that makes him the paragon of animals. It was intelligence and not brute force that enabled primitive men to overcome great beasts of prey, and it is intelligence joined with ethical ideals that alone can guarantee future progress. Emotional behavior is highly infectious; a dog fight sets all the dogs in the neighborhood into a frenzy; an excited chimpanzee will set a whole colony of apes raging; and we know only too well how the mob spirit may spread through a peaceful community, or war psychology sweep through an entire nation. The only safety for society and advancing civilization is in learning to control these animal passions by intelligence and reason.

Throughout the period of recorded human history there has been a notable growth of freedom not only from the rigors of nature but also from the tyrannies of men. Freedom from slavery of the body, mind and spirit has been bought at a great price through long centuries of conflict and martyrdom, and one of the amazing revelations of the past few years is the compliant way in which millions of people in Europe have surrendered all freedom not only in government but also in speech, press, thought and conscience on the order of dictators. Even in certain sciences, freedom of teaching and research has been restricted or prohibited, in spite of the fact that the advancement of science rests upon freedom to seek and test and proclaim the truth. Dictators seek to control men's thoughts as well as their bodies and so they attempt to dictate science, education and religion. But dictated education is usually propaganda, dictated history is often mythology, dictated science is pseudo-science. Free thought, free speech and free criticism are the life of science, yet at present these freedoms are stifled in certain great nations "with a cruelty more intense than anything western civilization has known in four hundred years."

In spite of a few notable exceptions it must be confessed that scientists did not win the freedom which they have generally enjoyed, and they have not been conspicuous in defending this freedom when it has been threatened. Perhaps they have lacked that confidence in absolute truth and that emotional exaltation that have led martyrs and heroes to welcome persecution and death in defense of their faith. Today as in former times it is the religious leaders who are most courageous in resisting tyranny. It was not science but religion and ethics that led Socrates to say to his accusers, "I will obey the god, rather than you." It was not science but religious conviction that led Milton to utter his noble defense of intellectual liberty, "Who ever knew truth put to the worst in a free and open encounter? For who knows not that truth is strong, next to the Almighty?" It was not science but religious patriotism that taught, "Resistance to tyrants is obedience to God." The spirit of science does not cultivate such heroism in the maintenance of freedom. The scientist realizes that his knowledge is relative and not absolute, he conceives it possible that he may be mistaken, and he is willing to wait in confidence that ultimately truth will prevail. Therefore, he has little inclination to suffer and die for his faith, but is willing to wait for the increase and diffusion of knowledge. But he knows better than others that the increase and diffusion of knowledge depend entirely upon freedom to search, experiment, criticize, proclaim. Without these freedoms there can be no science.

Science should be the supreme guardian of intellectual freedom, but in this world crisis only a few scientists have fought for intellectual freedom, and organized science in the countries most affected has done little or nothing to oppose tyranny. Science has flourished under a freedom which it has not created and it is sad to see that to-day, as in former centuries, it is left largely to religious bodies to defend freedom of thought and conscience, while great scientific organizations stand mute. I am proud of the fact that our own Association for the Advancement of Science adopted at its Boston meeting in 1933 a ringing Declaration of Intellectual Freedom.

The proposal was recently made in England that the British Association for the Advancement of Science and the American Association draft "a Magna Carta, a Declaration of Independence, proclaiming that freedom of research and of exchange of knowledge is essential, that science seeks the common good of all mankind and that 'national science' is a contradiction in terms." I am glad to report that those two great Associations for the Advancement of Science have for the past year or two been engaged in bringing about more intimate relations in the common tasks that confront all science.

We who are the inheritors of the tradition of liberty of thought, speech and press and who believe that freedom and responsibility are essential to all progress should use our utmost influence to see that intellectual freedom shall not perish from the earth. Such freedom has been essential for the advance of science, and the time has come when scientists and scientific organizations should stand for freedom.

There is no possibility that all men can be made alike in personality, nor any reason why all races and nations should hold the same political and social ideals. But there are grounds for hoping that they may come to cherish the same ethical concepts, for the needs and satisfactions, the instincts and emotions of all men are essentially similar. Upon this fact rather than upon uniform opinions, the hope of universal ethics rests. Science is everywhere the same in aims and methods, and this fact greatly strengthens the hope that in a world bound together by science into one neighborhood there may come to be common ideals regarding fundamental ethics.

The greatest problems that confront the human race are how to promote social cooperation; how to increase loyalty to truth, how to promote justice, and a spirit of brotherhood; how to expand ethics until it embraces all mankind. These are problems for science as well as for government, education and religion. Each of these agencies has its own proper functions to perform. Instead of working at cross purposes these greatest instruments of civilization should and must cooperate if any satisfactory solution is to be found. Scientists will unanimously agree that the spirit and aims and methods of science must be followed by all these agencies if any permanent progress is to be achieved; they will unanimously agree that science should cooperate to the fullest extent with government and education, but unfortunately there is no such unanimity of opinion when it comes to cooperation with religion. The memory of many conflicts between science and theology in the past and the knowledge of the existing antagonism of many religious bodies to science has generated a reciprocal antagonism on the part of many scientists to all religion. If the humanitarian aims of both science and religion could be viewed in the spirit of sweet reasonableness it would be seen that the differences between them are not such as to prevent fruitful cooperation in promoting human welfare.

Science as well as religion consists of both faith and works, principles and practice, ideals and their realization. The faith, ideals and ethics of science constitute a form of natural religion. Scientists generally would agree, I think, that the faith and ideals of science include the following: (1) Belief in the universality of that system of law and order known as nature. (2) Confidence that nature is intelligible and that by searching our knowledge of it may be increased. (3) Recognition of the fact that knowledge is relative, not absolute, and that only gradually do we arrive at truth concerning nature. (4) Realization that there is no way to avoid temporary error, since in unexplored fields we learn largely by trial and error. (5) The necessity of freedom, openmindedness and sincerity in seeking truth. (6) Confidence that truth is mighty and will prevail and that even unwelcome truth is better than cherished error. (7) Realization that truth can not be established by compulsion nor error permanently overcome by force. (8) Belief that the long course of evolution which has led to man and society, intelligence and ethics, is not finished, and that man can now take an intelligent part in his future progress. In these articles the faith of science does not differ essentially from that of enlightened religions.

The ethics of science regards the search for truth as one of the highest duties of man; it regards noble human character as the finest product of evolution; it considers the service of all mankind as the universal good; it teaches that both human nature and humane nurture may be improved, that reason may replace unreason, cooperation supplement competition and the progress of the human race through future ages be promoted by intelligence and good will.

In its practical aspects the ethics of science includes everything that concerns human welfare and social relations; it includes eugenics and all possible means of improving human heredity through the discovery and application of the principles of genetics; it is concerned with the population problem and the best means of attaining and maintaining an optimum population; it includes all those agencies which make for improved health and development, such as experimental biology and medicine, endocrinology, nutrition and child study; it includes the many scientific aspects of economics, politics and government; it is concerned especially with education of a kind that establishes habits of rational thinking, generous feeling and courageous doing. In spite of notable advances of our knowledge of these subjects we still know too little about human nature and the causes of social disorders. The extension of the methods of experimental science into this field is bound to be one of the major advances of the future. The ills of society, like the diseases of the body, have natural causes and they can be cured only by controlling those causes.

It is often charged that science is worldly, materialistic and lacking in high ideals. No doubt this is true of some scientists as it is also of some adherents of religion, but this is no just condemnation of either science or religion. Scientists as well as religionists have all the frailties of human nature and both fall short of their highest ideals. It has always been true and will continue to be true that knowledge outruns practice and that ideals are better than performance. Shakespeare said: "If to do were as easy as to know what were good to do, chapels had been churches and poor men's cottages princes' palaces." Or in the language of Mark Twain, "To be good is noble, but to tell others to be good is noble and no trouble." This is the age-long problem with which religion and ethics have struggled, namely, how can men be induced to live up to the best they know? How can they be brought to substitute the spirit of service for selfishness, love for hate, reason for unreason? The long efforts of past centuries show that there is no rapid solution of this great problem. But in the cooperation of science, education and religion there is hope for the future.

The American Association for the Advancement of Science is proud of its eighteen thousand members; the new edition of "American Men of Science" will contain nearly thirty thousand names. But the Christian churches of the United States number among their members about fifty-five millions. In so far as these churches represent the spirit of their founder they are concerned especially with the cultivation of ethics. That so little has been accomplished and so much remains to be done is due in part to refractory material, poor methods and the necessity of repeating this work in every generation. These religious bodies are enormous organizations with great potentialities for good. Why should not science and religion be allies rather than enemies in this process of domesticating and civilizing the wild beast in man?

The ethics of great scientists is essentially similar to that taught by great religious leaders. A scientist not friendly to organized religion has said that the Decalogue of Moses might be accepted as the Decalogue of Science if the word "Truth" were substituted for the word "God." Ivan Pavlov, the great Russian physiologist, left an ethical bequest to the scientific youth of his country, which reads like the warnings of the ancient prophets. Over the tomb of Pasteur in the Pasteur Institute in Paris are inscribed these words of his: "Happy is he who carries a God within him, an ideal of beauty to which he is obedient, an ideal of art, an ideal of science, an ideal of the fatherland, an ideal of the virtues of the Gospel." John Tyndall, no friend of the church, pronounced this eulogy of Michael Faraday, one of the greatest experimental scientists who ever lived; "The fairest traits of a character, sketched by Paul, found in him perfect illustration. For he was 'blameless, vigilant, sober, of good behavior, apt to teach, not given to filthy lucre.' I lay my poor garland on the grave of this Just and faithful Knight of God."

As scientists we are inheritors of a noble ethical tradition; we are the successors of men who loved truth and justice and their fellow-men more than fame or fortune or life itself. The profession of the scientist, like that of the educator or religious teacher, is essentially altruistic and should never be prostituted to unethical purposes. To us the inestimable privilege is given to add to the store of knowledge, to seek truth not only for truth's sake but also for humanity's sake, and to have a part in the greatest work of all time, namely, the further progress of the human race through the advancement of both science and ethics.

The Public Relations of Science

Wesley C. Mitchell
Columbus, 1939

Wesley Clair Mitchell (1874–1948), *economist, specialized in the theory of business cycles and national income during the first decades of the 20th century. Despite straitened family circumstances, he managed to attend the University of Chicago, where his interests in philosophy and economics were stimulated by courses under John Dewey, Thorstein Veblen, and William Hill. After a year at the universities of Halle and Vienna, he returned to Chicago, where he received the Ph.D. in 1899, summa cum laude. He taught economics there and at the University of California (Berkeley) before accepting a position at Columbia University in 1913.*

An originator of business cycle analysis in the United States, Mitchell helped to organize the National Bureau of Economic Research in 1920. He was also interested in the history of economics: his History of the Greenbacks *was a standard authority on inflation in the Civil War, and he was known for his lectures at Columbia on the history of economic theory. In addition to academic responsibilities, Mitchell donated many years to public service. He chaired the Social Science Research Council (1927 to 1930), was a member of Hoover's Research Committee on Social Trends, the National Planning Board, Works Progress Administration, and the National Resources Board. He served as president of the American Statistical Association, the American Economic Association, and the AAAS (1938).*

Until recently the attitude of the public toward science seemed to be growing more appreciative. There have always been folk who objected strenuously to the supposed implications of certain scientific hypotheses, but on the whole science was generally esteemed the most progressive factor in culture, man's best hope for bettering his lot upon earth. Of late this tide of approval has ebbed. There is a widespread disposition to hold science responsible for the ills men are bringing upon themselves—for technological unemployment, for the rise of autocracies, for the suppression of freedom, for the heightened horror of war. For their part, scientific men are appalled at the hideous uses to which their discoveries are put. They feel an urge to combat the misuses of science, to protect the social values they cherish, but what they can do is not clear. The quandary is one that all who cherish science should face, however unwelcome and difficult the task. I offer no apology for asking your attention to a discourse of uncertain issue on an unpleasant theme thrust upon us by developments we deplore.

I.

Let me start by recalling certain changes in the relations of science to society that may help us see our present problems in historical perspective.

The beginnings of scientific knowledge have been traced to man's dealings with the implements of his daily life—the sticks and stones, the skins, fibers and clay he shaped to his uses, and in the shaping learned to know. Human beings are born speculators; even the simplest cultures have their explanations of matters that puzzle us to-day—the diseases, weather changes, animal behavior, the creation of the world, what happens after death. Without this speculative bent human intelligence could not have evolved far; but neither could it win much useful knowledge without subjecting speculative explanations to practical tests. As the nursery of scientific thinking, the humdrum work of making and using household equipment had this great advantage: it required frequent repetitions under roughly similar conditions, when there was no great emotional stress, and when attention was centered upon immediate material results. In such activities it was least difficult to find out what operations were followed by the desired consequences, and what operations were superfluous.

Whether or not we accept this speculation about the humble beginnings of science, we know that at a later stage of cultural advance thinking about natural phenomena, like thinking about religious observances, tended to break away from direct associations with daily work. The slowly improving techniques of tracing the motions of heavenly bodies, keeping track of the seasons, measuring lands, developing mathematical propositions and erecting large structures called for unusual intelligence and training. Possessors of special knowledge wished to guard their trade secrets, to make mysteries of them, to initiate merely a few of their own choosing, and so increase their prestige. Mathematics is the subject least dependent upon the use of material objects, and it led the development of systematic thought, closely followed by its cousin, logic. Dealing with a rational universe of concepts, its affiliations seemed to be with religion and philosophy rather than with industry. So far as knowledge dissociated itself from technology, it escaped from continual subjection to matter-of-fact tests. That left it free to pursue attractive lines of speculation, but took away its most vigilant corrective and its sharpest spur to self-criticism. Even mathematical inquiry lost its momentum when it soared aloft in mystical flights.

Such efforts to understand the world as the Middle Ages made were concerned chiefly with problems of a divine dispensation. Observation was not pertinent, and factual tests of conclusions were not possible. The highest authority upon all questions was Holy Writ, which nobody might question and which the church interpreted. This orientation made acceptable the later Greek preoccupation with formal logic and disdain of matter. Aristotle, that great investigator, was transmuted into an obstacle to further investigation of mundane phenomena. Intellectual acumen achieved triumphs in its chosen fields but understanding of natural forces was not prominent among them.

The re-birth of science in the sixteenth and seventeenth centuries was brought about by turning from the study of concepts back to the study of nature. The new orientation was characterized by close observation, by the invention of devices to make observation more penetrating and accurate, by purposeful experimentation to simplify the processes observed, by close attention to quantity as well as to quality, by the practical application of mathematics to express the relations observed, by reformulation of concepts to fit the findings, by critical checking of one investigator's work by others, by the cumulation of tested conclusions in old fields of research and by the extension of this mode of inquiry to new fields. Inventing instruments for observing, setting up experiments, measuring and testing brought science again into intimate touch with the practical arts. Investigators took a keen interest in current affairs, sought to profit by the skill of craftsmen and to put what they learned to practical uses. Discoveries were applied not only to the production of goods, but also to navigation, fortification, ballistics and administration. By the close of the seventeenth century the dramatic achievements of "natural philosophy" were leading many to expect an almost limitless advance, and the promotion of science was recognized as a proper object of public policy. Kings lent their patronage to scientific societies. Philanthropists followed the royal precedents by offering prizes for improvements in the arts and later by endowing research.

Of course the public relations of science were not uniformly harmonious in this age of genius. But the celebrated clashes between scientific discoveries and beliefs held by churchmen did not affect many lines of inquiry and did not gravely retard the rising tide of investigation. Not less characteristic of the age than Galileo's troubles were Newton's services to churchly teachings and to the state. Scientific men have lamented that he devoted his later years to arguing the validity of biblical prophecies; they have paid less attention to his work as Master of the Mint. It was adjustments in the relative weight of the guinea and the shilling suggested by Newton that gave England a *de facto* gold standard in the eighteenth century, though Newton did not foresee this result.

An even more striking example of close relations between research and service to mankind is the life of Benjamin Franklin. The foremost American discoverer of his time, he was foremost also in applying and disseminating science to make life more comfortable, more secure, more interesting, more humane. These activities were incidents in the life of a busy printer, editor, politician, postmaster, legislator, colonial agent and diplomat. But while we wonder at the extraordinary versatility of a man who could become both a scientific discoverer and a great statesman under any conditions, we must remember that in Franklin's day science was still in its "natural philosophy" stage.

With the cumulation of results, science became a more exacting mistress, requiring of her votaries more exclusive attention. But science did not draw away from the material tasks of daily life as it did in Greece. On the contrary, these relations were becoming more intimate, while scientists were learning to speak symbolic dialects less and less intelligible to the public or even to one another. Let me illustrate the seeming paradox by the relations between science and industry.

To most of us the modern age is characterized by technological progress as markedly as by scientific discovery. We think of the two achievements as interdependent. This interdependence was less obvious to Franklin's contemporaries than to us. Theirs was a century of great inventions, but inventions made mostly by men not trained in science. The famous "agricultural improvers" worked by empirical methods. The great textile inventions came from handicraftsmen, one of them a barber. Metcalf, Telford and Macadam, the road builders, were "practical men"; so also was Brindley, the canal builder. Newcommen, "father of the steam-engine," was an ironmonger and blacksmith; his co-worker, Cawley, was a plumber and glazier. The Darbys, who found out how to smelt iron with coke, began as small ironmasters. A few inventors, it is true, tried methodically to discover scientific laws—Smeaton and Watt are eminent examples. Also, some scientists developed inventions out of their discoveries, as Franklin did with the lightning rod, and some set deliberately about the solving of industrial problems—Leblanc developed his process of making soda to win a prize offered by the Paris Academy of Sciences. But these instances were harbingers of a coming day rather than representative products of the eighteenth century.

This new day brought with it a division of labor in the conduct of industry matching the specialization evolving in science. The captains of industry who carried the Industrial Revolution through its youthful phases were often technical experts, business executives and capitalists united in one person. Men of this versatile type are still to be found even in "big business"; but they are becoming as rare as once they were common. For, as technology was elaborated, experts with special training were required to supervise its operations. Engineering became a learned profession—or rather a family of learned professions that multiplied by fission. It found a place in institutions of higher learning beside theology, law and medicine. Inventions continued to come from geniuses with little training, but more and more of them were made to order by experts. Business management meanwhile became so intricate, what with its problems of financing, selling, accounting, selecting personnel, planning investments, and the like, that a good-sized corporation required a staff of men with different skills to do part of the work that an old-fashioned captain of industry had performed for his small establishment. With a considerable lag behind engineering, business administration also raised claims to professional standing and developed schools of its own. As for capital, the requirements of business utilizing modern technology speedily outran the resources of the single enterprise or partnership. By a series of inventions not less important than those of mechanics, methods were developed for bringing together the savings of countless individuals, rich and poor, by providing types of securities well enough adapted to their several needs to attract funds.

With this double division of labor, in science and industry, the scientist could stick closely to research and feel confident that whatever applications of his discoveries were feasible would be taken in hand by men who knew more about industry than he. The engineer could devote himself as sedulously to technological matters, putting research problems up to laboratory workers and leaving business worries to executives. The latter could get technical experts of many sorts from the schools, and could expect cumulative improvements in technology from the joint labors of scientists and engineers. Investors frequently knew little about the enterprises for which they provided capital; they inclined to rely upon the advice of professional financiers and to protect themselves by spreading risks.

The economic results produced by this unplanned organization of mutually stimulating activities astonished mankind. Industry after industry reorganized its processes time and again to take advantage of the latest engineering applications of scientific discoveries, and new industries kept cropping up. The efficiency of human labor increased greatly, per capita income rose, and hours of labor declined. Higher standards of living and applications of science to the prevention and cure of disease reduced death rates and prolonged the average duration of life. Population grew rapidly in the nations that led the scientific procession, and spread where it would over the earth, dominating, exploiting, sometimes exterminating the non-scientific peoples. Life became ampler if not easier for the beneficiaries of science.

What industry owed science is repaid in many ways. It provided in bewildering variety laboratory equipment more accurate and powerful than that made by hand. It stood ready to construct any new contrivance an investigator designed, and often improved upon the original plans. Fortunes accumulated in business were the source of many scientific endowments. Business corporations granted research funds to universities, and set up research staffs of their own, which were sometimes permitted to work upon fundamental problems.

Governments recognized the social importance of science by making place for an expanding array of scientific courses in public schools and universities, and by undertaking wide-ranging programs of research. In this country, the Federal Government became the largest employer of scientific men. At the time of the Civil War it chartered the Academy of Sciences, and in the World War the National Research Council to advise it upon scientific problems; in 1934 it set up the organization that has developed into the National Resources Planning Board with affiliations covering the full gamut of the sciences.

Finally, the public at large had a share in these great changes. It was the ultimate beneficiary of reductions in costs of production, of increasing per capita output, of new types of consumers' goods, of shorter working hours, of better protection against disease, of free education. And this share was not wholly passive. Wage-earners adapted themselves with less friction than might have been expected to the working conditions imposed by the new technology. If many disliked the impersonal regimentation of the factory and the monotony of machine tending, others delighted in their control over stupendous forces, in the precision of the work they turned out, in the efficiency of the organizations of which they were parts. It is a grave mistake to overlook the enthusiasms evoked by machines, big and little. As recent experience in Russia reminds us, a population that has not become mechanically minded in large measure can not effectively use modern technology. And at home the masses welcomed factory-made goods. In successive generations they thanked engineering and science for illuminating gas, sanitary plumbing, kerosene, telephones, electric wiring, inoculation against epidemic diseases, automobiles, motion pictures and radios. Almost every one participated in scientific discoveries to the modest degree necessary for using these contrivances with some skill. Besides material products, many folk enjoyed what Tennyson called "the fairy tale of science"; their thoughts were pleasurably enlarged by the telescope and microscope. Pasteur, Edison, Mendel, Mme. Curie, Einstein became romantic personalities to tens of thousands, rivaling in popular appeal politicians, business leaders, actors and athletes. Without this eager welcome from society at large, government, business and schools could not have fostered research as they did, and science could not have progressed so rapidly.

It can not be said that these eminently cordial relations between science and the public were consciously engendered by scientific men. Now and then when some scientific hypothesis or procedure was attacked, individual scientists rushed to the defense and sometimes organized protests by scientific bodies. The controversies over what the public called "Darwinism" and "vivisection" are examples. Also scientists answered calls for their services freely and took advantage of opportunities to make their livings on lecture platforms, in schools, governmental bureaus and business enterprises. On appropriate occasions they dilated eloquently upon the service of science to civilization, and investigators with skilful pens and need of royalties wrote popular books. But most of the men who made modern science what it is devoted themselves single-mindedly to research. Their deeds, not their words, won the esteem and raised the hopes of mankind.

In short, this policy of *laissez-faire* worked wonders. Science helped industry, and industry helped science. Even the backward art of agriculture, which faces so many difficulties and uncertainties, was benefiting by research. The dreaded "law of diminishing returns" seemed to be overbalanced by improvements in practice based upon the work of soil chemists, botanists and geneticists. The frightful prospect of overpopulation that Malthus had taught the thoughtful to fear seemed to be dissipated by scientific agriculture and scientific techniques of contra-conception. Best of all, science seemed to have found the secret of illimitable progress. What it had done was merely an earnest of the greater things it would soon do. One discovery led to another so continuously that men began to take for granted a cumulative rise in the standard of living. They moved the Garden of Eden from its traditional place at the beginning of human history into the calculable future, dreaming of a world from which poverty would be banished. If some souls felt oppressed by the materialism of the age, youthful sages arose from the non-scientific peoples to argue, in the words of Hu Shih, that there is more of the spiritual in the scientific effort to control natural forces than in passive resignation to poverty and disease.

II.

I doubt that any scientist ever accepted without qualification this idyllic version of the benefits science confers upon mankind. Certainly there were numerous protests from scientific quarters against misuses of the new technology. Geologists and economists warned against the rapid depletion of mineral deposits. Chemists feared for the nitrogen content of the soil. Geographers and meteorologists protested that wholesale cutting of forests and the plowing of grass lands produced deserts. Biologists lamented the extinction of animal species and anthropologists the callous stamping out of simpler cultures. Social scientists found much amiss within the countries that were most progressive. Urban and rural slums persisted as centers of disease and crime. The need of securing capital to utilize the new technology put control over it into the hands of the propertied classes. Labor was often grievously exploited. Huge fixed investments that could be used for only one purpose made competition destructive. The obvious escape from these hazards was to form monopolistic combinations. That was pleasant for the monopolists but not for other business men or for consumers. Besides the obvious dangers of exploitation, many feared that the great combinations might purposely slow down technological advance because it threatened rapid obsolescence of their equipment. Business did not manage even its own interests properly, for every few years it generated a crisis and depression in which it suffered along with the whole community. And the international relations of the scientifically advanced peoples showed at his worst "the old savage in the new civilization." Demonstrations of the economic advantages of free trade no more stopped the imposition of protective tariffs than demonstrations of the horrors of war kept peace. Militant nationalism seemed to be spreading and growing more passionate. An appreciable fraction of scientific energy was devoted to contriving weapons of destruction. Thus against the glowing picture of science as a benefactor of mankind could be set a dark picture of science putting more power into the hands of certain individuals, classes, nations, generations, giving them a differential advantage over others which they exploited according to their several natures.

Though some of the Jeremiads I have been recalling belong to an earlier time, they did not produce a profound effect upon the public relations of science until recently. The ills complained of could be regarded as "growing pains." They were thought of as social "problems," which should be dealt with by arousing public opinion in a campaign of education that would lead to remedial legislation. Problems that could not be solved by this time-honored method would yield presumably to the slower processes of general enlightenment.

This optimistic attitude was particularly characteristic of democratic nations. It assumed tacitly that experts could devise whatever "reforms" were needed, and that the majority of voters were intelligent enough to understand and well disposed enough to support desirable changes. Science had a stellar role in this program for remedying the ills incidental to progress. It did not claim knowledge of good and evil but it enabled men to make their value judgments more intelligent by tracing the consequences of actions. Many people were devoting their energies to the study of social problems; they spoke optimistically of their subjects as social "sciences." It seemed not too much to hope that science might presently begin to guide social practice in somewhat the same fashion as it guided practice in industry and medicine.

A man might be skeptical of the nascent social sciences; but he could scarcely deny that the leading scientific nations managed to readjust their economic, political and social institutions when the new technology produced results they deemed bad. True, the readjustments were usually made by attacking one evil at a time, without due consideration of indirect consequences, which were often unfortunate in good part, and sometimes canceled the gains. Also the processes of reform lagged so far behind the changes produced by applications of scientific discoveries that new troubles began and new social adjustments were needed before the preceding reforms had been perfected. Despite all this, the scientific nations believed themselves to be evolving a social order adapted to the times—one that enabled them to grasp ever more fully the ever larger benefits scientific progress was bestowing upon mankind. I question whether history can show another period in which human hopes soared so high as in the closing decades of the century between Waterloo and the first World War.

III.

That the public relations of science have recently become disturbing both to the public and to scientists is due, not to any change in the character of science or the behavior of scientists, but to changes in social conditions. While most people approved on the whole the applications of science before 1914, they have come to dislike many of the effects produced by later applications. To be specific: when scientific improvements in one industry after another threw men out of work in earlier decades, the victims might suffer in silence or protest riotously and perhaps smash machines. But the public at large was not deeply concerned over their sufferings; it repressed disorder, expected the displaced men to find new jobs for themselves, and blessed science for reducing costs of production. Now that a larger part of the public suffers from loss of work or obsolescence of investments, science is blamed for technological unemployment. When the modern arts of communication were used to facilitate the political processes of democratic nations, they were extolled on all sides. Now that these arts, further improved, are controlled in some countries by autocratic governments and used to suppress opposition, many good people treat science as the culprit. When the scientific nations used their superior arms against backward peoples, only a few sensitive souls were wrathful over the unfairness or iniquity of the procedure. Most people felt that science was good when it gave them a decisive advantage over those they wished to "civilize." Now that these same nations are threatened by still more terrible weapons in the hands of their peers, their moral horror is sincere, and they wish scientific warfare back to the pit from which it was digged.

This shift in attitude toward science as one happens to benefit or suffer from its applications is doubtless a mark of human frailty, but it is one at which scientists should not cavil without recalling a similar frailty of their own. Now that we are on the defensive, we discover that science is neither good nor bad in itself, but is merely an instrument that can be put to good or bad uses, and that the blame for bad uses should be visited upon those responsible for them. But when science was being lauded for good works, who among us argued that the credit belonged, not to science, but to those who used it for the benefit of mankind?

We made this discovery when difficulties forced us to think more carefully about the place of science in society. Well as the old policy of *laissez-faire* in public relations worked for a time, it had encouraged in us an indolent complacency foreign to the critical spirit of inquiry. We may not enjoy the shocks that have aroused us any more than an investigator rejoices over facts that disprove an elegant hypothesis; but we must face the situation and see what we can do to mend it. Every one concerned with the future of science or of mankind bears a share in the responsibility for trying to understand the present situation, and to decide what action, if any, is called for. It is not likely that a satisfactory solution can be produced in short order, for the problem is one of numerous variables in shifting combinations. But I venture to suggest an obvious proposition that seems to me of controlling importance, and to point out certain corollaries that should guide both our attempts to understand the public relations of science and our future policy concerning them.

IV.

The fundamental proposition is that scientific research is a social process as much as business, political or religious activities are, and as such is interwoven with all other social processes, influencing them and being influenced by them.

My historical sketch of the public relations of science supports this view. But let me borrow two illustrations from thoughtful physicists. David L. Watson has shown that social institutions impose their pattern upon research, putting a premium upon conventional inquiries and obstructing originality. Like other social activities requiring close cooperation, research gets organized, organizations become bureaucratic, bureaucrats may have routine efficiencies, but originality of thought and cordial welcome to originality in others are not conspicuous among them. Watson compiles a formidable list of instances in which scientific organizations have been slow to recognize fundamental discoveries, and he makes his point more uncomfortable by naming certain rather odd contemporaries who may be doing work more important than the men to whom we award medals.

In a way that should come close to home to every investigator, P. W. Bridgman has pictured the intellectual struggles a scientist must undergo if he strives seriously to live "an intelligently well-ordered life." Accustomed to subject his concepts in physics to operational tests, Bridgman tried to treat with similar rigor the concepts that count most in social intercourse. The results were disconcerting.

> Not one of our social institutions [he found] rests on the
> secure foundation that we so easily assume when we refute the
> skeptic or instruct our young. Never has any institution been
> justified in terms that anyone capable of close thinking could
> accept without stultification. Yet if ever the tragic need for close
> thinking and intelligent convictions on social questions was
> obvious it is at the present.

This loose thinking that characterizes social intercourse introduces confusion into the lives of all, though the confusion may be recognized only by those who try to live intelligently. For every one derives the words in which he thinks from society. They bring into mental processes all sorts of implications inherited from the past that will not bear analysis, and from which one can free oneself only by laborious analysis of the sort that has made the exact language of physics differ from the ambiguous language of everyday life.

Scientific research, then, is one among many social activities carried on by the peoples of our culture. Like all such processes, it is carried on by men who learn in childhood languages ill-suited to close thinking; by men who wish to eat, to make love, to win approval as well as to know; by men who are reared in an environment of emotional likes and dislikes; by men who become so absorbed in their technical tasks that they have little energy to criticize the non-scientific parts of their own make-up. And these scientific men form a tiny fraction of their communities. So far as they succeed in emancipating themselves from the misconceptions and prejudices prevailing in their social groups, they become by virtue of their partial emancipation queer creatures whose judgment most people mistrust outside of their specialties. Both the temperament that inclined them to research and the habits they form in research tend to make them awkward, ineffective, reluctant in appealing to the emotions that are so potent in influencing men. It is difficult to see how a few scattered individuals, each accustomed to think for himself and to be critical even of his fellow inquirers, can guide public opinion except by slow educational processes. In the long run their thinking may rule the world, just because it serves the purposes of mankind better than the traditional thinking it gradually replaces. But in the short run, others take of scientific discoveries only the parts that have an immediate application, and put these parts to such uses as they see fit—uses that serve whatever aims these others pursue. The prompt and potent influence of science upon society comes from these uses, good and bad, which scientists control only in small part.

Even in democratic countries, then, scientific men find it hard to bridge the gulf between their attitudes and those of the general public. In autocratic states the governments might give scientists fuller opportunities to direct public policies than they enjoy in democracies. But the autocratic states known to us are not built on that model. They are avid for science, to be sure, but only for science that is an uncritical servitor of ends the rulers determine. As between the difficult public relations confronting them in democracies and the shackling of free inquiry confronting them in autocracies, scientists can not hesitate. Theirs is a world of intellectual freedom, not perfect, alas, but the freest world the mind of man has yet created, and to let any authority under any pretense prescribe what conclusions they shall accept as scientific is to stultify the spirit of science.

V.

What, then, can scientists do to improve their public relations in communities where they are relatively free?

As I see the situation, they have two sets of opportunities and responsibilities: first, their opportunities and responsibilities as citizens; second, their opportunities and responsibilities as investigators.

It must be admitted that to many scientific men the performance of civic duties is an unwelcome interruption to their research work. Some brilliant investigators are temperamentally unfitted to share in the processes by which a democracy reaches its decisions. Among the great discoverers of the past there have been cynics who despised the common herd, recluses who could scarcely endure social contacts, geniuses so erratic that their judgments upon practical affairs seemed crazy, rationalizers who urged the insignificance of one citizen among millions as an excuse for shirking responsibilities. Presumably representatives of these types exist among our scientific contemporaries; but I know no ground for supposing that they form a larger proportion of the persons listed in Dr. Cattell's "American Men of Science" than of those listed in "Who's Who in America." A goodly majority of scientific men have normally balanced personalities and are competent citizens. They can be counted upon to take their civic responsibilities as conscientiously on the average as any other group, and to act with as much common sense.

What scientific men can do as citizens is like what other intelligent men can do. If democracy is to work well, many people must form considered judgments upon a wide variety of problems. In forming a considered judgment on a given issue, what experts have to say should be taken into account. Who these experts are depends upon the character of the issue; more often than not contributions are needed from several kinds of specialists. All the many species of the genus scientist belong at one time or another in the list of desirable technical advisers; so also do lawyers, business organizers, labor leaders, social workers, educators, civil servants, politicians, and so on. When matters within the competence of some group of scientists are involved, they should contribute what they know, whether formally invited to do so or not. To make their advice effective they should welcome help from people more skilled than themselves in the arts of popular presentation. On matters concerning which a scientist has no special knowledge, he should listen to others and form the best judgment he can from what they advise. To an individual this task of sifting and weighing different opinions is time-consuming and difficult. On complicated issues organization is needed to bring into focus all the intelligence available in the community. Hence, one of the civic duties incumbent upon all scientific men in common with other citizens is to support vigorously but critically the nascent movement toward organizing all the intelligence we possess for constructive study of social problems before they become pressing emergencies that have to be dealt with in a hurry that allows no time for careful thinking.

The outside limits of what scientists can accomplish as citizens are set by their ignorance. Not merely does no individual have more than a tiny fraction of the knowledge that is needed; all the scientists of the country put together do not know enough to solve many of the problems that a democracy faces. In addition to the responsibilities they share with all other citizens, scientific men have the special duty of trying to increase the kind of knowledge required to deal intelligently with public problems. Their opportunities and responsibilities as citizens merge into their opportunities and responsibilities as investigators.

From the social view-point, the most urgent item in the unfinished business of science is to increase knowledge of human behavior. If we had keener insight into individual psychology, we might not be able to alter fundamental drives, but we might be able to direct them into beneficent channels. Preaching righteousness doubtless prevents men from being as bestial as they might otherwise become. Appeals to reason prevent them from making as many errors as they otherwise might. But the moralist and the rationalist admit that the results of their efforts are grievously disappointing. Scientific men with any gift of self-analysis realize that they have their own shares of selfishness and animosities. To subdue traits in oneself is hard enough to give an inkling of the difficulty of controlling them in society at large. Perhaps, and perhaps is all we can say, if we can come to a clearer understanding of how we behave, we can learn how to condition men so that their energies will go less into making one another miserable.

One of the things we have learned about individual behavior is that it is influenced greatly by social environment. In John Dewey's phrase, "all psychology is social psychology." Improving knowledge of social organization and its working is therefore part and parcel of the urgent task of learning how men behave. Though we may believe ourselves citizens of the most fortunate nation in the world, we have no more reason for complacency about the way in which our social organization works than for complacency about individual behavior. For example, our economic organization does not permit us to buy from one another as much wealth as our workers are able and eager to produce. Even in the best of years we fail to provide a national income large enough to give American families on the average what experts on household economics hold to be a standard of living adequate to maintain efficiency. In bad years this inadequate income falls off by a fifth or a sixth; in the very worst years by 40 per cent or more. All this is true of our industrial equipment and practice as they stand. Proud as they are of our technological progress, engineers know that much of our equipment and many of our methods are far behind the times. We fail to make full use of knowledge that technological applications of scientific discoveries have put at our disposal. I might develop the shortcomings of our economic organization at great length, and then go on to exploit the weaknesses of our political and social institutions. It is needless to do so; for every candid and intelligent citizen can point out defects, however convinced he may be that with all its faults, the American scheme of institutions is the best in the world. If scientists can do more than other intelligent citizens toward improving social organization, their contribution will consist in raising knowledge of social practice.

We all know that the social sciences lag far behind the natural sciences. That is because they deal with phenomena more complicated, more variable and less susceptible of experimental manipulation. Since investigators can not experiment at will upon social groups, they can not effectively apply to their problems the methods that have made the laboratory sciences strong. Max Planck once told J. M. Keynes that in early life he thought of studying economics but found it too difficult. Of course economic theory as we have it to-day is far easier to master than physical theory. But Planck was a true scientist, one who wished to gain knowledge that accounted for actual phenomena. He had learned enough to realize that it is far harder to get such knowledge of economic than of physical processes.

Yet the case of economics and its sister sciences is not hopeless. The rapid growth of statistics is providing mass observations upon social behavior of many kinds; the equally rapid growth of statistical technique enables us to learn more from a given array of data than our predecessors could. These materials and methods are making it possible to measure many social factors, some rather accurately, some roughly. Uniformities appear not only in averages but also in the way in which individual items are distributed about their means. Statements in terms of probability can be substituted for vague statements about the effect a certain cause "tends" to produce. True, work on this observational basis encounters many difficulties. It is limited by the variety, extent and accuracy of reliable data upon human behavior. It is laborious, slow and expensive. In presenting his work a realistic investigator begins with a critique of his data and methods, he ends by setting forth the probable errors and limitations of his results, and the road from the beginning to the end may be long. Instead of definitive conclusions he thinks others should accept, he presents tentative approximations he expects others to improve. The work has not even the advantage of calling for less hard thinking than speculative theorizing; for the relations among the variables in the problem are seldom manifest of themselves. All that can be claimed for this type of work is that it deals with actual experience, that its results stand or fall by the test of conformity to fact, and that it grows cumulatively after the fashion of the observational sciences. But that is enough to give mankind strong reason for following this lead in seeking the knowledge required to improve social organization.

I do not imply that the social sciences can rapidly become such assured guides to social progress as the natural sciences are to technology. Because of difficulties inhering in their subject-matter, the social sciences will continue indefinitely to lag behind the natural sciences in precision and reliability. For a long time to come we shall have to form our opinions on many social issues in the light of common sense rather than of science. Knowledge of past experience should prove helpful in this uncertain process, and advice from specialists who have studied this experience should be sought. But wise technical advisers in these difficult matters will not pretend to certitude. As citizens we shall do well to suspect the intelligence, the candor or the disinterestedness of those who promise sure cures for social ills.

Scientific men are wont to face facts, whether these facts conform to their wishes or not. Most of them are sufficiently emancipated from conventional thinking to look critically upon social institutions. They contrast the society of today with its poverty in the midst of plenty, its class, racial and international animosities, its puerile aims and its destructive methods, to a society they can imagine living in security and comfort, using its increasing knowledge to provide a finer life for all mankind. This contrast should not be accepted in a spirit of resignation. It is a call to action. But scientific men will not be true to their own standards, they will not render to society the largest service of which they are capable, if they let their actions be guided by their feelings. No current discouragement should blind us to the great strides in human welfare made since science assumed its modern form; no fit of impatience over delays and relapses should make us forget that knowledge is won step by step, through the toilsome efforts of thousands of men. To jump this work with its numberless failures and its gradually cumulating successes, expecting to land in Utopia, is to give up faith in science for faith in magic. Men who take scientific methods seriously as the best hope of floundering mankind will seek to apply them just as critically and remorselessly in their social as in their physical thinking.

But science can not flourish in the future and yield the fruits for which we hope unless freedom of thought prevails. That is a condition we have been inclined to take for granted as part of the heritage our predecessors won. Now we realize that what they fought to win we must fight to maintain. The investigator's right to follow truth wherever it led was part of the common man's right to freedom of conscience and freedom of speech. These rights were established by political struggles and embodied in political institutions. The democratic way of life and the scientific way of thinking grew up together, each nourishing the other. If one now fails the other will falter. Where democracy is suppressed today science is fettered; for autocracy can not brook disinterested criticism of its dogmas or its practices. Freedom of scientific work in the years to come can be guaranteed only by preserving the institutions that secure freedom to all citizens. Perhaps scientific men have more at stake than any other social group in the struggle to maintain democracy.

To this struggle they can make a crucial contribution. The fate of free societies hangs upon the wisdom or folly of mass decisions. The gravest dangers to democracy come from within, not from without. They are ignorance and propaganda that turns ignorance to its uses. The best way of dispelling ignorance is by diffusing knowledge. The most effective defence against meretricious propaganda is critical inquiry. John Dewey is warranted in saying that "the future of democracy is allied with spread of the scientific attitude." To foster this attitude among their fellow citizens by all means within their power is a duty incumbent upon us who cherish science. As teachers in schools and colleges we can help thousands to develop respect for evidence. As citizens we can be brave opponents of prejudice and hysteria. We can promote general understanding of the methods and results of science through our own writings or those of allies more skilled in popular exposition. These things we should do, not as high priests assured that they are always right, but as workers who have learned a method of treating problems that wins cumulative successes, and who would like to share that method with others.

World War II was often called "the physicists' war." Long before the atomic bomb made the claim a modest one, American leaders were insisting that a hundred physicists were "worth a million soldiers." Mobilized more effectively than ever before, and on an undreamed-of scale, the scientific community quickly assumed an imposing central position in the popular mind and in military and governmental circles. "Science cities" were assembled at Oak Ridge, Los Alamos, and Hanford; science "empires" grew and were consolidated at the California Institute of Technology, the University of California at Berkeley, the University of Chicago, and MIT. The dramatic end of the war with Japan impressed the new importance of science—especially physics—into everyone's consciousness.

Even before the United States entered the war, President Roosevelt had created the Office of Scientific Research and Development, which provided the framework for the mobilization of scientific talent. The men behind the office were its director, Vannevar Bush, a former vice-president of MIT, chairman of the National Advisory Committee for Aeronautics, and then president of the Carnegie Institution of Washington; Karl Compton of MIT; James B. Conant of Harvard; and Frank Jewett of Bell Laboratories. Toward the end of the war, Bush was asked by President Roosevelt to report to him on the possibilities of a scientific mobilization during peacetime. In his letter to Bush, Roosevelt wrote glowingly of the uses of science for the nation's health, more jobs, and the betterment of the standard of living; "New frontiers of the mind are before us."

Bush's report, *Science—the Endless Frontier*, has become a classic programmatic statement. In it, Bush insisted that "without scientific progress no amount of achievement in other directions can insure our health, prosperity and security as a nation in the modern world." Arthur Holly Compton's address of 1944 likewise reflects the renewed confidence and growing sense of destiny of the scientific community. "[S]cience is not only a servant," he insisted, "it also gives orders." Compton predicted that, because science now provides the foundation for military might, "it is an absolute 'must' for a nation that would maintain its place in a warlike world that it shall keep its science in the front rank." As scientists, Compton concluded, "it is our primary task to give our country the strong foundation of science necessary for her proper growth. If we can also find for ourselves the way to useful and satisfactory citizenship in the society built on that foundation of science, the world will follow our leadership."

If the war and the Bomb made the physicist into an awesome public creature, both feared and admired, they also drew him into the public arena in a way his forebears had warned about. Vast new bureaucratic enterprises to handle government-sponsored research were created. The Atomic Energy Commission, the National Science Foundation, the National Institutes of Health, the Office of Naval Research—all were creations of the postwar period.

Science required, and got, vast sums of money for research and development. According to National Science Foundation figures, government expenditures on research and development in 1940 amounted to $74 million (0.8 percent of total budget outlays); by 1949, the figure had increased to $1,082 million (2.7 percent of total budget outlays). For the 1951–52 research and development budget, the Department of Defense provided 53 percent and the Atomic Energy Commission 36 percent. Basic research was always dwarfed in the total research and development budget by applied research and by development. But it should be noted that, by 1955, for *basic research alone*, the Atomic Energy Commission and the Defense Department were two of the largest customers, far outdistancing the National Science Foundation. Scientific leaders recognized that, unlike the period after World War I, when science's service to industry was stressed, now the United States was to operate on a footing of semipermanent war. Characteristic of the Cold War period that followed World War II were the close intellectual and financial ties among the military, the quasi-civilian Atomic Energy Commission, and the universities. The universities operated a system of national laboratories in conjunction with the Atomic Energy Commission; individual scientists drew project support (and their universities drew substantial overhead) from Army, Navy, Air Force, Atomic Energy Commission, National Institutes of Health, or other government agencies. Characteristic also of the period was the rise of independent and only semiprivate research agencies—the Institute for Defense Analyses, The Johns Hopkins Applied Physics Laboratory, the RAND Corporation, the Center for Naval Analyses, and others.

Kirtley Mather called attention to these interconnections in his 1952 address. While noting that politicians need scientists and will increasingly require their services, he stated that scientists likewise have become dependent upon public support, and "beyond doubt the scientists will be requesting increased support from the politicians in the coming years." Mather called for an end to the mutual suspicions between them. He urged those scientists with courage to "take the calculated risk of cultivating the common ground of science and politics."

He was quite right, but, in 1952, his statement required considerable courage. The Cold War, while swelling science budgets, also curdled political discourse. The early 1950's were known even then as the McCarthy

Era. When the fear of communism became a national paranoia, Mather rather forthrightly blew the whistle on it. "Political orthodoxy rather than mere technical competence," he charged, "has been accepted as a basic qualification in many academic institutions and industrial laboratories, even when the work is completely unclassified and does not involve access to anything that could be considered a military secret." The most famous casualty of the hysteria was, of course, J. R. Oppenheimer, but there were others.

The foreign-born had special difficulties. American science was enriched with refugees from fascism in the 1930's and 1940's, but after the passage of the McCarran Act in 1950 the State Department severely restricted the influx of foreign scientists and impeded the international exchange of information and service at U.S. universities by foreign-born scientists. "Certainly the scientists of America cannot be expected to do their best work," Mather maintained, "as long as they remain in the stultifying atmosphere that has been imposed upon them by political trends."

An interesting contrast to Compton's address is that of Warren Weaver. During a period when the omnipotence of science was being sold in many sectors of American life, Weaver warned that "superstitions" about science were unhealthy, not only for the layman, but for the scientist as well. He cautioned against the widespread image of the scientific-technical priesthood; instead, he saw "science as the servant of man, not his master; and as a friendly companion of art and of moral philosophy." Like Mather, Weaver was concerned about the climate of fear that had seriously damaged the openness of science. "Who will save us," he concluded, "but the scientist and the humanitarian. Yes, the humanitarian in science, and the scientist in the humanity of man."

The year 1957 was a critical one for the scientific enterprise. It was the year of Sputnik and the beginning of accelerated competition with the Soviet Union. The biggest customers of research and development, the Department of Defense and the Atomic Energy Commission, would be joined during the next decade by the National Aeronautics and Space Administration. For science, it was the start of a bout of space fever. In his December 1957 address, Paul B. Sears relayed his concern about the skewing of the scientific enterprise:

> [W]e have looked upon science as an expedient rather than as a source of enlightenment. To be specific, our very proper concern with the applications of mathematics, physics, and chemistry may be clouding the fact that we need biology in general and ecology in particular to illuminate man's relation to his environment.

In a sense, this address laid bare concerns that in ten years' time would become openly critical. Behind the bigger science budgets and the demands for more scientists and increased science education lurked a deep insecurity. "We are beginning to sense," Sears claimed, "that the elaborate technology to which we are so thoroughly committed makes us peculiarly vulnerable." In an era in which security was the password to public largesse, Sears concluded that "[o]ur future security may depend less upon priority in exploring outer space than upon our wisdom in managing the space in which we live."

What Science Requires of the New World

Arthur H. Compton
[no meeting] 1944

Arthur Holly Compton (1892–1962) *was a physicist whose achievements encompassed the fields of science, industry, education, and public service. The Compton family could boast a distinguished record of service to science and education: his father was a professor of philosophy at the College of Wooster; his oldest brother, Karl, a scientist and president of the Massachusetts Institute of Technology, held numerous government-related advisory positions; a third brother, Wilson, was president of Washington State College.*

After graduation from Wooster (B.S., 1913), Compton took up advanced studies at Princeton University, receiving the Ph.D. in 1916. For the next three years he taught physics, worked as a research scientist for Westinghouse, and developed aircraft instruments for the Signal Corps. One of the first recipients of the newly established National Research Council fellowships in 1919, Compton spent a year at the Cavendish Laboratory at Cambridge University with Ernest Rutherford. Here he began a series of experiments on the scattering of x-rays which he continued at Washington University upon his return to the United States. The Compton effect (the increase of wavelength of scattered x-rays) grew out of these investigations. He received the Nobel Prize for physics in 1927.

As professor of physics at the University of Chicago (1923 to 1945), Compton turned to the theory of ferromagnetism and to cosmic radiation in the 1930's. An active member of the National Academy of Sciences, Compton initiated in 1941 an Academy study of atomic energy's military prospects and later became director of the Metallurgical Laboratory of the Manhattan Project, where Fermi and his team developed the first controlled nuclear chain reaction. Compton was president of the AAAS in 1942.

. . . Once more, because of the rigors of war, we have found it impossible to hold the annual meeting that has been our tradition for almost a century. My own colleagues, as typical members of our association, are this afternoon in their laboratories, engaged as devotedly as any member of the armed forces in the effort to preserve our country's freedom. Yet the world comes to us as representatives of science with searching questions. We must pause to give a considered answer. "This is a war of science and technology," they tell us. "Do the forces of freedom have the knowledge, skill and technical resources needed to bring victory?" "After the war is over how will science have changed our world?" The nation asks us, "What of the night, and what of the day that is to dawn?"

Unconditional answers to these questions can not be given. Yet it is possible to say something about the present balance of scientific power and to point the direction in which science makes it necessary for the world to move.

I have accordingly chosen as my subject for today, "What Science Requires of the New World." For science is not only a servant; it also gives orders. There is a legend that Daedalus, the Greek hero who first learned how to work with steel, toiled long and hard with his forge and anvil to fashion a sword. This he presented to King Minas to replace his old one made of bronze. The citizens of Crete came to him in consternation. "This sword will not bring us happiness," they complained, "it will bring us strife."

"It is not my purpose to make you happy," replied Daedalus. "I will make you great."

Science is the steel of Daedalus. His sword is the weapons we are fashioning with the aid of science. As his steel, so likewise science brings new social conflicts and changes in treasured traditions. But also like the steel of Daedalus, science is compelling man to follow the sure road to a greater destiny. Let us try to catch a glimpse of this new world that science bids us enter. Can we attain a stable peace? In the post-war world, what changes in our mode of life does science demand?

First, let us note that science affords real hope for a stable peace. We find ourselves in a world conflict with the mighty powers of science strengthening the arms of both contestants. "A hundred physicists in this war are worth a million soldiers." This oft-quoted statement was made by one of our leaders early in the struggle. Consideration of the military significance of radio, magnetic mines, methods of submarine detection and a variety of new weapons which the physicists have initiated or developed indicates that this estimate may not be greatly exaggerated. With equal justification, however, one might have singled out the vital contributions to the war effort made by research in chemistry, mathematics, immunology, aerodynamics or some other field of science and technology. We have learned that knowledge is strength, and that intensive scientific research is the only way of supplying certain types of knowledge that are essential to waging modern war.

The fact is that science and technology are now spending extraordinary efforts on supplying means of destruction and methods of protection against attack. At first it was our enemies who had the scientific advantage. This was because for a prolonged period the Axis powers had been making extensive scientific as well as military preparations for the war. There is evidence, however, that in most fields of military operations our technical advances are now coming more rapidly than are the enemy's. When added to our overall industrial advantage and our superiority in supply of materials and men, if this scientific advantage can also be maintained, there is little room for doubt of the success of our armies.

We do not discount our opponent's strength. We know the high quality and resourcefulness of his scientific and technical men. We accordingly expect reverses as well as successes. Yet the Allied Nations are in a superior position with regard not only to availability of materials, industrial capacity and numbers of fighting men, but also with regard to scientific and technical strength. Because we have this strength, and are determined to preserve our freedom, we may lay our plans for the future on the assumption of victory.

When peace has then been won, can the world be kept stable?

Let us assume that the United Nations have gained a complete victory. All indications are that the world will still be actively war-minded. However successful may be our armies and the efforts of the negotiators of peace, a great conflict such as the present war will leave many wrongs unrighted and large groups of people resentful at their fate and filled with fear and hatred toward their neighbors. Weapons of destruction are being developed of hitherto unheard-of power. No one can consider the armadas of mighty bombers, with flying range to reach any target, and the increasing amount of destructive bombs they carry, without fear of a yet more disastrous war to come. There will be a nation smarting under the restrictions placed upon it and ambitious for power. What will prevent it from welding these weapons into a war machine with which it will snatch the mastery of the world?

The only answer to this threat is preparedness and vigilance by the powers in control. Preparedness and vigilance have always been the price of continued peace. What science and technology have brought into the picture is a change in the type of precautions that must be taken. Improving the chance for stable peace is the increasing time and magnitude of the preparations required to wage successfully a modern war.

Consider what is needed to exploit the power of airplanes. Here is a weapon whose development and production strains the technical resources of the greatest nation. The same is true of the factories that would build tanks or the laboratories that would develop electronic devices superior to those of an ingenious and highly skilled enemy. It is one of the great safeguards of the stability of modern society that the weapons with which wars are now won are the product of cooperative research and manufacture on so large a scale that the effort can not be hid. If precautions to maintain peace are to be taken, we must assume the establishment of a world policing system with power to learn what nations are doing that may constitute hazards to the public safety, and determination to stop unlicensed building or accumulation of arms. The large scale of modern military preparations makes such policing much more practicable than was true before technology became the basis of military power.

But, we are asked, can not some new weapon be developed secretly on a small scale which is nevertheless so powerful that those who hold it will have the world at their mercy?

Here again the trends of modern science give us considerable assurance. More and more the major inventions and industrial developments are the result of the cooperative efforts of large groups of research men. An idea may emerge in the mind of a lone inventor, but it passes through many hands before it is ready for use. Also, parallel developments by competing groups are the rule. It is rare indeed that a completed new industrial development catches the world by surprise. If this is true in industry, it is yet more unlikely that the balance of the world's military power will be upset by an idea kept secret from a vigilant enemy. To be of conclusive military significance not only must the idea be perfected; the new weapon must also be adapted to industrial production and be manufactured on a major scale. But such a development is difficult to hide. In spite of airplanes and radios, the world is still a big place, and a weapon that would conquer the world must be ready for widespread use. If we are alert we should know of any new military development of this kind before it has become a hazard to nations organized to protect the public safety.

One of the most necessary aspects of vigilance is the active cultivation of science. Not only is science the foundation for present military developments; it is also the means of opening up of new possibilities. It is an absolute "must" for a nation that would maintain its place in a warlike world that it shall keep its science in the front rank. The possibilities of present and new ideas of military import must be explored to be sure no competing nation will gain the advantage of being first in the field. Only by maintaining an active body of scientists can the foundation be laid for a strong military structure when it is required. Stability and peace in the new world thus can not be ensured unless the dominant world power keeps up a vigorous and continued growth of science.

Equally essential to military superiority is industrial strength, factories accustomed to doing large tasks rapidly, sources of raw materials and good communications. These are the tools that will make effective use of the ideas developed in the laboratories. Possession of large armies, navies and air forces are needed to start a war, but if the struggle is prolonged mere accumulation of such forces can not bring victory. Ultimate fighting power depends rather upon knowledge and facilities for building this knowledge rapidly into weapons as best fitted to changing conditions. Careful attention must thus be given to both the scientific and the technological foundation of military might.

The net result is that with the use of a world police force of a feasible size, it should be possible for a dominant governing body to maintain a stable peace in the new world that science builds. In spite of mighty new weapons that may tempt some ambitious leader to try again to snatch control of the world, the need for vast scientific and technical development to produce such weapons gives a stability adequate for a vigilant governing power to keep the peace.

So much for the *negative side*. We have seen that in the world that science is shaping an alert government should be able to prevent serious wars if only it will maintain the strong science and technology that are the basis of modern military strength. What now of the *positive side*?

Scientific men are becoming increasingly conscious of their social responsibilities. They begin to realize more clearly the tremendous human implications of the forces which their investigations are introducing. A parent is eager that his child shall contribute something worthwhile to society. So the scientist is eager that his science shall work for human welfare. He sees vast new possibilities for betterment of life, and he is impatient to see these possibilities become realities. More and more those concerned with science are endeavoring to ensure the wide use of the products of science.

But I am not concerned today with what we as members of the American Association for the Advancement of Science may *want* our sciences to do for humanity. I would call attention rather to those changes in society which growing science and technology make inevitable. I am not referring to new gadgets or improved standards of living, nor even to better health and longer life. These are the obvious and direct results of applied science. We know they will come as science continues its task, and they are welcome gifts, indeed. But I am thinking rather of the inescapable trends that follow the principles of evolution. Only those features of society can survive which adapt men to life under the conditions of growing science and technology.

There are three such features which I shall use as examples. These are (1) increasing cooperation, (2) better and more widespread training and education, and (3) rise of commonly accepted objectives toward which society will strive. Note that such changes mean growth to greater manhood. That science makes them inevitable was what Daedalus meant when he said his steel would make men great.

Finding ourselves then in a stable new world, how will the conditions of life be changed?

Perhaps the most significant change in the life built by science and technology will be the increased organization of people into larger groups concerned with performing common tasks. People will become yet more specialized, and consequently will be increasingly dependent upon each other.

H. G. Wells has called attention to a remarkable example of evolution, in which during the short period of a thousand generations an organism has been observed to change from an individualistic animal like a cat to a social animal like a bee or ant. He refers, of course, to man, who twenty-five thousand years ago lived in caves with his loosely bound family, and now lives in vast communities in which each works for the group and depends upon the other for most of his means of living. Here is evolution in progress—social evolution, if you please—man becoming civilized.

My present interest in this process is that the chief forces bringing about the socialization of mankind have always been those of increasing knowledge and techniques, of which the characteristic present-day representatives are science and technology. For thousands of years each village has had its butcher and baker and candlestick maker, specialists in their trades who supply others with their wares. The scientific age has greatly increased this specialization. It is not enough now to be a chemist or an engineer; one becomes an organic chemist specializing on long chain esters, or an electrical engineer specializing on echoes of short electric waves. The nation needs this special knowledge, and supports the few who have it by the efforts of millions of others. In fact, the world itself is not too large a unit to support and use effectively the work of such specialists.

Were it not for technology, by which the work of each person is greatly multiplied through use of power machines and methods of mass production, the work of many highly specialized individuals could not be supported. Were it not for science, which has made possible such developments as steam engines, airplanes and the radio, there would be no markets of continental extent which absorb technology's mass production. Combining the specialization of science and the mass production of technology, a society is built of unparalleled richness and strength.

A noteworthy feature of this modern society is that its strength depends to an unprecedented extent upon the cooperation of its members. Since the specialist can live only through the help of others, cooperation is the corollary of specialization. We have learned that the automobile demands sobriety, and that congested life in a city requires careful attention to sanitation. Similarly, we are now learning that life in the world built by science and technology is possible only with widespread cooperation.

In the fight for survival by various forms of society one sees the evolutionary process operating in full force. During times of stress weaker societies are absorbed or replaced by stronger ones. Cooperation thus becomes essential to the survival of any social régime.

Effective means of securing cooperation accordingly becomes a major objective of political systems. Breakdown of cooperation in the ranks of the enemy becomes a most useful method of weakening an enemy in wartime. Replacing a monarchy with a democracy, introducing the military rule of fascist dictators, rallying a nation to support of a communistic state, all are examples of attempts to secure a more effective basis for cooperation in the common tasks of society.

The first essential in securing widespread cooperation is to develop a widespread desire of people to work together. Several methods of securing such will to cooperate are effective. The most certain is to present people with a common danger, such as attack by an enemy. It is this that has built a nation out of the Chinese people, and that has made the strength of our nation grow beyond our dreams during the present conflict. Another powerful method is to present the group with an inspiring objective. Thus Hitler calls to the Germans to make of themselves a master race, Lincoln challenges his countrymen to strive "that freedom shall not perish from the earth." With such an ideal men and women lose themselves in working for the common cause.

But also in the more prosaic periods of normal, peaceful life the cooperativeness of a community depends upon having common interests. If specialized science requires cooperation, cooperation in a society of free people requires the will to work for the common welfare. "Without vision the people perish" applies with tenfold force to the modern world. Whether this vision comes from the loyalty bred of a common danger, from political or economic expedience, from philosophical principles or from the inspiration of religious teaching, the will must be there. Otherwise the inhabitants of a specialized community can not obtain their needs without conflict, and the great advantages of technological society have turned into tragic liabilities. Science thus requires of the new world that its people shall want to work together for the common good.

Let us note further that the extent of the social unit in which this cooperation occurs is being increased rapidly by science. Typical of the forces working in the direction of expansion is the radio. Its music, stories and news are heard over large areas. The radio advertisements make possible the sale of products over a market of continental size. As a result the optimum size of a strong political or economic unit is rapidly growing. Eighty years ago our country was almost split asunder by divided interests. Now its continental size gives great advantages in both industry and government, and the divisive forces are lost amid the concern with mutual trade and the technical and military strength that comes because the country is big. Science itself is as extensive in its interests as the human race. So also is religion. Nations may try to make themselves self-sufficient, but this is in the interest of security as opposed to an advantageous economy. Economic considerations alone would make trade extend freely throughout the globe.

The advantages of such global extensiveness are evident in the British empire, and to lesser extent in our own country. A phone call to a distant city brings at once the information needed to complete a design. Tools for an emergency repair are flown across the continent. An international misunderstanding is quickly cleared by flying a statesman ten thousand miles across the sea. Pictures and stories and songs representing the life of one people become familiar to all the world. Such are the forces with which science is drawing the world together into one great community. The fate of the world from here on is a common fate.

It is the American men and women of science who are pioneers in shaping this world community. In no other part of the world is life so dependent upon the products of science as among ourselves. This is clearly demonstrable in terms of the number of kilowatts or of radios or of tons of steel or gallons of gas or telephones per capita. People in other parts of the world look with dread to the time when their lives will be altered by technology. It will upset their ancient customs. Things long cherished will disappear and be replaced by things new and strange. We do not dread science and technology; it is natural to us as a part of our lives. Of necessity the world must look to America as the pioneer in finding how to live a satisfying life in a society based on science.

Amid the standardized products of mass production, how are the ultimate values of individual life to be attained? How can we best cultivate the spirit of mutual helpfulness so highly important for a satisfactory life in a technological régime? What type of education will fit citizens for a useful and satisfying life in such a world? We in America face these questions first and in their most acute form. We who represent American science are those who have perhaps had most first-hand experience in hunting for the answers. The world looks to us to point the way.

One thing we have already found is that technology and science place unprecedented value on education. The use of steam and electric power has decreased the need for common labor, while growing specialization has increased the need for those who coordinate our activities. Skilled labor, however, remains vital to American society for building and operating our machines, and is rewarded with shortened hours and higher pay. Business requires middlemen to handle its varied commerce. Vastly increased numbers of professional men and women have been absorbed in occupations of responsibility which before the era of technology were hardly known. Here we find the engineer, the secretary, the economist, the patent lawyer, the research scientist and many others. Those responsible for planning the work of society have never been so driven by ceaseless demands as in today's America. Reflections of this pressure are to be seen in the multiplication of governmental offices, in the rise of schools of business and public administration and in the growth of the army's staff of supervising officers. The masters of society have indeed become the servants of all, in an unresting labor that knows no release. By emphasizing the need for intelligent direction, and reducing the need for unskilled labor, technology is thus spurring Americans of all levels toward an ever higher standard of training and education.

Most significant, however, of the factors that make life worth while is the vision of a goal that one recognizes as worthy of his supreme effort. Now in wartime Americans find that goal in the victory that will preserve our freedom. When peace comes, what will be the objective that will unite our efforts? Will we be inspired by the new possibilities presented by science for making the world suitable for the highest needs of man? Here is a challenge of a millennium that science presents to religion. For is it not the great task of religion to show us the goals for which we should strive?

But whether we call it religion or humanism or social expediency, acceptable objectives must and will be found. This follows again from the fact that the will to cooperation needs a challenging purpose, and cooperation is essential to the survival of a social system. If we fail to develop adequate objectives our society will be replaced by another that has such objectives. It was Hitler's call to the youth of Germany to lose themselves in the greatness of the German Volk that gave such strength to what had been a sick nation. It was Lenin's challenge of a great new society based on equality of all and the glory of work for the common welfare that has made of modern Russia a mighty power. To Americans the values inherent in freedom had been almost forgotten. We were weak from lack of objective, until the Japanese attack united us to meet a common danger.

Perhaps the great objective for us will be that of the common welfare, as discovered by a hundred million citizens who become educated to the possibilities of common men. Leaders we have, and new ones must arise, who will give us the inspiration of great ideals.

As scientists, it is our primary task to give our country the strong foundation of science necessary for her proper growth. If we can also find for ourselves the way to useful and satisfying citizenship in the society built on that foundation of science, the world will follow our leadership.

To Daedalus, steel was much more than metal for fashioning swords. It was the means of making men grow to greatness. So likewise science.

We have in science a powerful weapon with which to fight our war for freedom. If the powers in control will be vigilant and will establish a suitable policing system, science and technology are giving us a world in which a stable peace can probably be maintained. But science requires changes in our mode of life. The specialization of our society based on science must be matched by ever closer cooperation on a rapidly increasing scale. Growing attention to special training and more extensive education for leadership is inevitable. Such developments give promise of a truly great society. We are, however, in need of the inspiration of a commonly accepted social objective that will unite our willing efforts.

Never has man had so real an opportunity to master his own destiny. With the new ideas of science, the new tools of technology and the new view of man's place in nature that science has opened, we see ever more clearly how we can shape our world. May God grant us a vision of the possibilities of man which will challenge us to the worthy use of these great new powers.

The Common Ground of Science and Politics

Kirtley F. Mather
St. Louis, 1952

Kirtley Fletcher Mather (1888–) *became interested in geology in high school, and pursued the subject at the University of Chicago and Denison University in Ohio (B.S., 1909). He began graduate work at Chicago, but interrupted his studies for several years of teaching and field work with the U.S. Geological Survey in Arkansas and Oklahoma. The experience provided data for his dissertation on invertebrate fossils and began his long-standing affiliation with the Survey.*

After receiving his Ph.D. from Chicago in 1915, Mather took academic positions at Queen's University, Ontario, at Denison in 1918, and at Harvard University from 1924 until his retirement as emeritus professor of geology in 1954. As Survey geologist, Mather spent summers doing field work on the glacial geology of Colorado and Alaska, and the oil fields of the Midwest. His expertise in petroleum geology was recognized by private companies, which engaged him in surveys in Bolivia, Argentina, and Nova Scotia.

Early in his scientific career, Mather saw the need for popular scientific education, and he broadcast a series of lectures that became a popular book, Old Mother Earth *(1928). Mather was particularly concerned with continuing education for adults, organizing a Center for Adult Education in Boston in 1933, and served as director of Harvard's summer school from 1934 to 1941. Mather was president of the AAAS in 1951.*

It is my purpose to speak along lines forecast by the healthy tradition that has lately been developed by my predecessors in this particular spot in the Annual Meeting of the American Association for the Advancement of Science. Instead of discussing some specific problem or some notable progress within the field of geology, my special department of the physical and biological sciences, I shall consider certain of the broad problems arising from the impact of science on modern life. . . .

At first thought, there are those who might cynically inquire, "Is there any common ground at all between two such antagonistic fields of activity as science and politics?" Conflicts between scientists and politicians have been so widely publicized in recent months that there might seem to be adequate basis for such a question. Scientists have criticized politicians for their ignorance of the strategy and procedures that have proved so efficient

113

in the progress of science. Politicians have berated scientists for their impractical idealism and have even denounced them as subversive when they object to security regulations and procedures that seem to them inimical to the continuing development of scientific knowledge.

It is obvious, however, that America needs both wise and honest politicians and intelligent and conscientious scientists. Either vocation should deserve as much respect and receive as much honor as the other. In spite of the fact that there was no reference whatever to science in the platforms of either the Republican or the Democratic party, drafted in preparation for the 1952 political campaign, the politicians who phrased those platforms were well aware that the successful candidates would necessarily depend heavily upon the work of scientists for progress toward many of the objectives they promised to reach.

Scientific research and development have become a major enterprise in the United States during the years of our lives. Politicians recognize quite generally the value of that enterprise and the fact that national security has become increasingly dependent upon its success. The strength of our military defense is measured by the achievements of science more than by any other factor. This is not just a matter of atomic bombs, whether of the fission or the fusion type. Of the three billion dollars spent on science in America during the years 1941–45, exclusive of the money devoted to the development of the atomic bomb, 86 per cent was paid by the U.S. government. Five-sixths of the federal expenditure was made through the Department of Defense.

Governmental agencies also depend upon science for the improvement of the public welfare, as well as for the strengthening of military defense. The federal budget for scientific research that contributes to public health, better living conditions, more efficient and widespread use of natural resources, improved educational techniques, and all such factors that make for "better living" is now well over half a billion dollars per year. It is quite unnecessary to labor the point. Politicians need scientists in their business, and will need them more rather than less, as the years come and go.

But this is a two-way street. Scientists are increasingly dependent upon politicians. Approximately 35,000 specialists in the physical, biological, agricultural, and engineering sciences are now employed in government laboratories and research facilities, working under the supervision of politicians. Much of their work is in the development and application of science rather than in fundamental research, but this is "science" nonetheless. Furthermore, thousands of scientists employed in colleges, universities, and industries are actually supported by federal funds, made available for that purpose by vote of the Congress. More than a hundred million dollars is channeled each year through educational institutions, and thrice that amount through industrial laboratories, for research and development under governmental supervision.

It was the scientists of the country who appealed to a reluctant Congress for legislation establishing the National Science Foundation. They had pointed out the many important functions that such an agency, supported by federal funds and administered by civilians, could best perform. The needs were obvious, the procedures clearly defined, and yet the response has not been nearly as generous as the nation's welfare demands. Beyond doubt the scientists will be requesting increased support from the politicians in the coming years.

Surely, all will agree that, if the interdependence of science and politics is not as clearly comprehended as it ought to be, something should be done to make their common ground more obvious. The time is certainly at hand for a moratorium on mutual recrimination, suspicion, and jealousy between scientists and politicians, and for a rebirth of a spirit of fair play, constructive cooperation, and mutual understanding. To that end, it is well to consider first the objectives that they hold in common.

Science has a dual objective, and therefore scientists have two functions. On the one hand, science is the quest for knowledge of a certain kind—the kind that leads to the formulation of general laws connecting a number of particular facts. Therefore, the scientist may seek knowledge solely for its own sake and for the satisfaction of his inner urge toward broader, deeper understanding of the nature of the world and of man. Such has been the motivation of many of the greatest scientists of all times. No consideration need be given to the consequences of the results of his search, insofar as they may be useful for any purpose whatever in the practical affairs of everyday life. The only requirement is that each newly acquired bit of information and each new conceptual scheme be fruitful for additional steps that he may take in his quest for knowledge and for truth.

On the other hand, science is power of a certain kind—the kind that makes it possible for men to manipulate nature. The scientist may thus seek scientific knowledge in order that it may be applied to increase the efficiency and the comfort of mankind. This may involve just as fundamental research as any other, or it may be merely the development of techniques and gadgets whereby new concepts are translated into practical operations. It is significant that when scientists today are philosophizing, they are more likely to distinguish between "fundamental research" and "technological development" than between "pure science" and "applied science." The fact is, of course, that every item of scientific knowledge ever gained, in response to whatever motive, has been found sooner or later to have practical significance, either directly or indirectly, in human affairs. Rash indeed would be the scientist who dogmatically asserted that knowledge about nuclear fusion could never contribute to human welfare, impossible though it may now seem to be to imagine that the hydrogen bomb could ever be anything but a horrendous weapon for cataclysmic destruction. Even that, of course, has its far-reaching significance to modern man.

Politics likewise has a dual objective. On the one hand, politics is the administration of the state and the management of public affairs, directed toward the maintenance of order within the nation and its protection against foreign foes. Therefore, politicians enact laws to govern the activities of citizens, limiting their behavior to orderly procedures, and to provide police power to restrain lawbreakers and protect the state. Army, Navy, and Air Force, in the political structure of the United States are, appropriately, administered in the Department of Defense. Concentrating on this objective of politics, the politician seeks power *over* people. Orders are issued; obedience is demanded; disobedience is punished. On occasion, the military power of the state is used for aggressive warfare, the better to protect the state from its external enemies, or for the purpose of bringing more people and more resources under its control.

On the other hand, politics is the administration of the state and the management of public affairs, directed toward the betterment of human welfare and the enrichment of the lives of individuals of whatever status. This far transcends the other objective, with its concentration on order and stability. In recent years, every state, the world around, has become increasingly a social service state. Politicians have found it polite to seek and gain power *with* people, rather than merely to seize and hold power *over* people.

There is also another bifurcation within the political sphere. Whether or not the greater emphasis is placed on one or the other of the two objectives of politics, the goals may be sought in either of two ways. One way involves government of all the people, for some of the people, by a very few of the people. That leads to regimentation within dictatorships and establishes an autocracy. The other way leads to government of all the people, for all the people, by all the people. That involves the universal acceptance of the responsibilities of self-government and establishes a democracy. To those who are prone to quibble about the fact that the United States is officially a republic rather than a democracy, I would suggest that our republic has the structure of a representative democracy, one of several possible alternatives for the democratic pattern of government.

Autocratic politicians, attempting to concentrate power in their own hands, are very likely to look upon science as a most important servant of the state. This, of course, is the attitude in the Soviet Union at the present time. The physical and biological sciences are employed to yield technological improvements, raise living standards, and strengthen police power. The social sciences are especially useful in maintaining control in the hands of the rulers. The techniques whereby minds are manipulated through propaganda have been described and in some instances developed by social psychologists. These techniques have been quickly learned and effectively used by politicians to maintain the supremacy of the state and to achieve their own ambitions for personal power.

To an even greater extent, the politicians in a democracy are necessarily concerned with the progressive development of the sciences and their application to all aspects of life. Such politicians are at least in theory the servants of the people. The citizens who have chosen them as their representatives demand that in all considerations of administrative policy high priority be given to their own economic, social, and cultural welfare. They are likely to insist that the benefits accruing from scientific and technologic achievements be shared as widely and as equably as possible. Both politics and science are expected to operate in the service of mankind. On that common ground the politicians and the scientists should work together in wholehearted cooperation.

Unfortunately, however, such cooperation is not now being displayed in the United States to anything like the extent that is obviously desirable. Competent scientists are reluctant to accept employment with the federal government, and many of those now in the civil service would be glad to leave it for other jobs. A survey of the attitudes of scientists toward various types of employment, made by the President's Scientific Research Board and published in 1947, indicated that of all the scientists questioned "only 11 per cent preferred a Government career. Thirty-one per cent preferred industry and 48 per cent the university environment. The remaining ten per cent preferred consulting work or some other activity." That low estimate of government service as a career is evidently not significantly a matter of financial rewards. The salary scales in industrial laboratories are far higher than those in government bureaus, but the latter are at least as remunerative as those in educational institutions.

That survey was made prior to October 1947. Since then, the search for disloyalty among government employees and the procedures followed by security boards have had their deleterious effects upon government employment. It is almost certain that a similar survey made today would reveal an even lower appraisal of the opportunity afforded by the government for a successful career in science, in comparison to that available in other ways.

Friction between scientists and politicians extends, even more unfortunately, far beyond the area in which the politicians hold the purse strings and therefore can enact the detailed regulations which the scientists must obey. The *Internal Security Act* of 1950, popularly known as the McCarran Act, and the *McCarran-Walter Immigration Act* of 1952 have dropped a "red tape curtain" around the United States, which in many evil ways resembles the Iron Curtain around the Soviet Union. Each of those measures was passed by the Congress over the veto of President Truman. In his veto messages, the president spelled out the harmful consequences of the legislation to the nation, and displayed a far clearer comprehension of the bad effect it would have upon science in America than was displayed by those who voted to enact it. The dire impact of that legislation upon science in America is so well known that I do not need to tell the story here.

This whole question of the conflict between intellectual freedom, essential to the uninterrupted progress of science, and national security, essential to the preservation of our country in this period of real danger, ought to have much more careful study than it has yet received. It is so confused by prejudice, suspicion, and fear that it is almost impossible to remove it from the fires of emotion and weigh it on the balance of reason. But unless this is done the welfare of our country will be seriously jeopardized.

Most of the freedoms that we hold so dear are relative freedoms, to be exercised only within more or less clearly defined limits. Some of them have to be abandoned or more sharply restricted in time of war, either hot or cold. The most basic freedom of all is intellectual freedom, the right of a man to think his own thoughts and announce them without fear to those who will listen to him. This is the freedom that is safeguarded by our Constitution's Bill of Rights, although there it is spelled out in terms of freedom of speech, of the press, of peaceful assembly, and of the free exercise of religion. It is, in fact, the very touchstone of democracy, the creator and preserver of the orderly flexibility that makes democracy so much more efficient and desirable than any autocracy. The real test of democracy is not applied by asking questions about the statements embodied in a nation's constitution or the presence or absence of ballot boxes and universal suffrage. If anyone wants to know whether the community, state, or nation in which he resides is truly democratic, let him ask this question: What actually happens to the member of an unpopular minority when he dares to speak his mind in opposition to the spokesmen of the popularity majority?

When that test is applied to the organizations and communities of scientists in the United States, they are found to fall within the democratic bands of the broad spectrum that ranges from stark autocracy at one end to perfect democracy at the other. In fact, many of the significant new ideas that have led to progress in each sector along the expanding frontier of science have been first proposed by an individual, or a small minority, in opposition to views widely held by large majorities. The novel concepts have been appraised in the market place of public opinion. Each scientist has been encouraged to form and express his own independent judgment. No hierarchy of academicians has decreed what is orthodox, or branded as subversive anyone who deviated from the approved "line." Even though one scientist may strongly disagree with another's opinions, he knows he must defend the other's right to express them, else he will be false to his calling as a seeker for more accurate understanding of the ways of nature. If the suspicion should enter his mind that perhaps in times of ideological conflict a little thought control might be desirable, he has only to remind himself of the sorry plight of the biological sciences in the Soviet Union.

Intellectual freedom for scientists inevitably conflicts with the necessity for national security. To what extent and in what ways should it be limited? The answer to that question has thus far been given by the politicians, with woefully inadequate consideration of the scientist's point of view. Political screening is necessary in certain sensitive areas where scientists deal with military secrets. Unfortunately, those areas have been either too loosely or too broadly defined. They should be restricted to the absolute minimum. The ideas of competent scientists concerning what that minimum should be ought to have far more respectful consideration than they have thus far received.

The nature of the political tests also needs reexamination. The fundamental difficulty here arises from a basic disparity between the mental processes of scientists and politicians. In debates across a frontier, the primary aim of the true scientist is to understand rather than to refute. In such debates, most politicians aim to demonstrate their own worth and insist upon the correctness of their own views, rather than attempt to understand an opponent's ideas. Altogether too few politicians in America display the intellectual qualities of political scientists.

Consequently, the scientist who tries to understand the motives and the behavior of people on the other side of the current ideological conflict is engaging in an intellectual enterprise quite foreign to the politician's mental and emotional habits. He is therefore open to suspicion and will almost certainly be caught by the political screen of "reasonable doubt." There is of course some truth in the well-known saying that a man is known by the company he keeps. But that method of appraisal is valid only when it is taken to mean: a man is known by all the company he keeps. To base conclusions solely upon a man's associations with a few organizations or individuals, especially selected by angry politicians, is both unscientific and unjust. Either all his associations, or none, should be considered by those responsible for political screening to ensure national security.

It would be bad enough if the harmful policies and practices were applied only to the scientists working on projects that have security implications. Actually they are extended far beyond that relatively small group. Political orthodoxy, rather than mere technical competence, has been accepted as a basic qualification in many academic institutions and industrial laboratories, even when the work is completely unclassified and does not involve access to anything that could be considered a military secret. Administrators dare not risk charges that might be made by Congressional committees, or by radio and newspaper commentators, that they are employing "red" scientists. Many institutions have their own security officers, who are concerned not only with the personnel employed in classified projects, but also with those engaged in scientific research not covered by security regulations. Particularly where academic tenure is not a stumbling block, it has been comparatively easy to dismiss, or to bar from employment, capable scientists accused of past association with organizations now considered questionable or subversive. The number of such tragedies is far greater than any statistician can ever discover. Both the unfortunate victims and the institutions by which they have been employed have characteristically shunned publicity, for reasons that may differ from case to case but add up to a definitely hush-hush policy.

Intellectual freedom involves the free interchange of information and ideas among scientists not only within our country but also between and among those of all nationalities. Among the most stimulating factors in scientific progress are the international gatherings of specialists in the various scientific disciplines and the visits of foreign scientists to the centers of research and development in the United States. Few persons, other than the scientists themselves, are aware of the tremendous indebtedness of American technology to scientific research prosecuted in other countries. Freedom to travel may not be one of the justly celebrated "Four Freedoms," but for the man of science it ranks at least as high as any other.

Especially since the passage of the *McCarran Act* in 1950, the Department of State has been exercising increasingly rigid control over the movements of American and foreign scientists, both in and out of our country. The power to withhold passports and visas is one which, when improperly used, can deal a telling blow to scientific progress. International gatherings of scientists have been curtailed, both in America and abroad, by the establishment of political and ideological tests of fitness to enter or leave the United States. Individuals of unquestionably high scientific competence have been denied passports for travel abroad, in some instances to pursue research projects of vital importance to our country. Foreign scientists of equally fine standing in their professions have been barred from entrance, even when invited for specific services to be rendered at some of our most important universities or laboratories. In each instance the refusal is explained on the grounds that the issuance of the requested passport or visa is not in "the best interests of the United States."

The repercussion of such events throughout the great body of American scientists cannot but be deleterious in the extreme. Not knowing what alleged crimes have brought this penalty upon one of their number, the scientists, upon whom the future security and prosperity of our country depend, find their efficiency decreased in subtle but significant ways. Each will decide that the only way to avoid a similar fate is to withdraw completely from all participation in public affairs. No longer will they dare to give critical consideration to the social implications of their work. In one large compartment of their lives, in which they should accept their responsibilities as citizens in a democracy, they will fit their minds into the Procrustean bed of rigid conformity to official governmental policies and majority opinion. How soon this acceptance of authoritarian dictates will carry over into the other compartment of their lives, in which scientific habits of thought have proved so invigorating for scientific progress, is a neat question for the psychologists. At the moment, it is generally held that schizophrenia is quite bad for man.

Certainly the scientists of America cannot be expected to do their best work as long as they remain in the stultifying atmosphere that has been imposed upon them by political trends. The appeal to get them out into the fresh air, where the winds of freedom and confidence will once more stimulate them to high intellectual adventure, is based not so much upon selfish desire for the personal welfare of individual scientists, as upon the recognition of what is absolutely essential to the continuing health of science as an important contributor to the future of America.

Scientists and politicians have so much to do when they get together on the common ground of service to mankind, that friendly cooperation, in mutual trust, should certainly be the aim of all. Scientists have solved most of the technical problems of reducing infant mortality and prolonging life through sanitation and medicine. Politicians have provided public hospitals and public health agencies. They are jointly responsible for the unprecedented increase in world population that has characterized the first half of the twentieth century. Scientists must now accept the responsibility of solving the technical problems involved in providing adequate subsistence for growing populations. And politicians must arrange for appropriate sharing both of technical skills and of the means of subsistence in all parts of the world.

These are not impossible tasks. I have no space to develop details— I simply assert that in my opinion the resources of the earth, when made available by the science of today and tomorrow, will prove adequate to meet all the needs of mankind for untold years to come. The present rate of population increase is both unprecedented and temporary. It is far lower in those parts of the world that are now enjoying an economy of plenty than in those parts still laboring within an economy of scarcity. The politicians of India have followed the advice of scientists and have adopted a decreased birth rate as an item of national policy. There is good reason to expect that the world's net reproductive rate will be appreciably lower by the end of the century than it is at present.

The goal for this joint task force of scientists and politicians should be a standard of living everywhere at a level adequate to promote social and political stability in a free and democratic society. That is quite different from attempting to provide an American diet and an American standard of living for all people everywhere. The physical and biological scientists are already well on the way toward solving the technical problems along the route to that goal. Their part of the task is in accord with their well-established habits of mind and their attitudes toward human welfare. For politicians, however, this will be a different story. The management of public affairs is inherently more difficult than the management of soils, seed crops, or livestock. Politicians will need all the help that the social scientists and the educators can give them. Moreover, politicians have for the most part been thinking within the framework of an economy of scarcity, even in lands where the industrial and agricultural revolutions, made possible by science and technology, have brought prosperity to fortunate people. It will be quite a feat of mental and emotional gymnastics for them to reverse the field and formulate policies appropriate to an economy of plenty.

Education will obviously play a major role in this project, and educators also are cultivating this same "common ground." Indeed, there is plenty of room within its broad horizon for artists, men of religion, and many others. In a very real sense, an educator is a sort of hybrid individual, combining many of the characteristics, methods, and aims of the scientist and the politician. Be that as it may, the educators, too, will have to blaze some new trails.

In many insufficiently developed regions, fundamental education, urgently needed by adults as well as children, should have much more emphasis upon science than is currently given. Adherence to long-established policies of teaching classics, literature, and philosophy leads to the slighting or even the omission of science in certain countries. Emphasis upon trades and technical skills, in other programs, fails to inculcate knowledge of the methodology of science.

Even in countries like our own, where the teaching of science is stressed almost or quite to the point of slighting the humanities and the fine arts, there is a challenge to educators far more basic than that of perfecting the techniques of education. The great body of scientific knowledge appears to be a collection of separate fragments. The conceptual schemes developed in the separate areas of investigative analysis seem to bear little or no relation to each other. Specialists in each narrowly prescribed field of research have their own vocabulary and speak a language that can scarcely be comprehended by others. The need for synthesis and integration among the sciences is urgent.

It is not enough for geologists to borrow techniques and apparatus from the physicists in order to engage in geophysical research. Or for biologists to equip themselves with the paraphernalia of the chemist's laboratory and announce that they are biochemists. Synthesis and integration must be sought at a much deeper level. Truly integrative science will seek basic concepts that are valid in all the scientific compartments, conceptual schemes that tie together the disparate knowledge now displayed in academic showcases. Above all, it will make more clearly visible the unity of the universe and the fundamental nature of its orderliness.

This is especially important for men who want to live in a free society. Chaotic aggregations, although susceptible to statistical description, leave the individual unit under the domination of capriciousness. Freedom is a function of order, not of lawlessness. As it reveals the fundamental place of order in the universe, integrative science will validate the intuitive urge of man for freedom.

The politicians of America have a vital interest here. Educational procedures in our academic system have all too often stressed philosophically unrelated technical specialization. Practical competence has been inculcated, without an accompanying emphasis upon culture. There has been premature dependence upon science alone, and ethical values have been sadly neglected. All this has led to confusion, irresolution, demagogism, and the crippling of the democratic body politic. Science without conscience is in danger of leading us all to catastrophe.

Scientists also have a stake in the area of integrative education. Not only must they accept responsibility for research along basic conceptual lines, but they must assist in the development of a climate of public opinion that recognizes the supreme worth of intellectual freedom. They must resist every effort to have education serve a society that maintains the symbols of political democracy, but actually concedes final control to covert concentrations of power that use educational skills to arrogate to themselves the administration of national welfare. Perhaps psychologists and psychiatrists might make their most valuable contributions in the next few years if they would concentrate their research, not on how to influence and manipulate people, but on how to free people from the compulsion to control others.

Less than one half of one per cent of the inhabitants of the United States are professionally engaged in scientific research, the technological application of its results, and the teaching of science. These scientists share the responsibility for human welfare with all other citizens. They cannot, however, escape the fact that because of their intellectual powers and their influence in forming public opinion, theirs is a larger share of that responsibility than their numbers alone would indicate.

Only those scientists who have both a social conscience and a large measure of courage will take the calculated risk of cultivating the common ground of science and politics. Knowing my colleagues as I do, I am confident that there are enough of them to make a powerful and beneficent impact upon public affairs in this country of ours, which even now has some claim to be "the land of the free and the home of the brave."

Science and People

Warren Weaver
Atlanta, 1955

Warren Weaver (1894–) *is an internationally known mathematician, science administrator, and humanist. He took degrees in mathematics and civil engineering at the University of Wisconsin (1917). A brief period of teaching mathematics at Throop College (California Institute of Technology) was interrupted by war service as a U.S. Army Air Service second lieutenant. Weaver returned to Wisconsin in 1920 and received a Ph.D. in mathematical physics for his work in electromagnetic theory. He remained at Wisconsin for 12 years, as professor of mathematics and, later, department chairman.*

In 1932, Weaver became director of the Natural Sciences Division of the Rockefeller Foundation, a philanthropic fund that had recently undergone a reorganization broadening the scope of its encouragement of scientific research. Although trained in physical science, Weaver proposed long-range support of quantitative biology when the foundation sought new directions for its science program. The foundation agreed, and in the period between 1933 and 1959 provided grants to the life sciences totaling $90 million, which Weaver believes was central to the development of molecular biology. After reaching the Rockefeller Foundation's formal retirement age in 1959, Weaver joined the Sloan Foundation as vice-president and, since 1964, as science consultant.

An important facet of Weaver's work has been a continued effort to interpret the scientific enterprise to the general public; for these endeavors in 1965 he received UNESCO's Kalinga Prize and the Arches of Science Award for outstanding contributions to the popularization of science. Weaver was president of the AAAS in 1954.

Because I feel so deeply and so strongly concerning what I have to say on the subject of science and people, I shall run the risk of being dully pedagogical and state my plan at once. First, I am going to ask what successes man has had in his various endeavors and inquire why science seems to bulk so large among these successes. I am going to recount some foolish ideas concerning science that have arisen partly because of its successes. I am going to contrast these with a series of statements that seem to me more accurately to describe science and its relation to life. The main conclusion will be that science belongs to all the people, and that this fact presents the American Association for the Advancement of Science with a great opportunity and a great duty.

Think of the various major tasks to which men have, over the ages, addressed themselves. They have sought food, warmth, shelter, and other guards against the physical assaults of nature. Each individual or group has also sought protection against attack from the rest of mankind.

Men have tried to understand the physical universe. They have striven to apply this body of understanding to attain control of and to exploit this power over physical nature.

Men have tried to understand organic nature—how it evolved, and how individual organisms reproduce, grow, and function. They have sought health of body. They have tried to understand the nature of mind, of consciousness, of memory, of the learning process. They have endeavored to manage personal relationships within family groups, the village, the tribe, the state, the nation, and the world at large. They have attempted social and eventually political organization at all levels of inclusiveness and complexity, and they have tried to understand human behavior as it affects all these interrelationships.

Men have created methods for ownership of property and have elaborated systems of customs and laws in an attempt to protect individuals and serve society. They have recorded history and have attempted to understand it. They have, at great cost and with high dedication, tried to strike a balance between regulation and liberty.

Men have sought to enrich life through development of the pictorial arts, literature, music, drama and the dance. They have created systems of logic and metaphysics and have tried to analyze the nature of knowledge and reality. They have formulated codes of esthetics and morals and have contemplated the purpose and meaning of life.

In this vast and interrelated range of concerns and activities, where do the successes lie? What things have men really done well?

Each man is entitled to his own answer, but my own reply would go as follows. Probably the most conspicuous, the most universally recognized, and the most widely applied success lies in the understanding and control of the forces of physical nature. Coupled with this, I would place the progress that has been made—even though it is but a start—in the understanding of organic nature.

But along with these two I would want to bracket, without attempting to suggest an order of importance, two other major successes. The first of this second pair of successes is to be found in the grandeur and practicality of the principles of personal conduct that have been enunciated by the great religious leaders. I would suggest, for example, that the Ten Commandments, the Golden Rule, and the rest of Sermon on the Mount have the generality within their realms that Newton's laws of motion have in theirs, plus the fact that no religious Einstein has found it necessary to insert correction terms of higher order.

The second further success that seems of major proportion is to be found in the degree to which life can be and has been enriched by the arts.

Thus it is my own conviction that the poet has done a job that science must thoroughly respect, and perhaps should envy.

In listing only these four major successes, some real unfairness may have been done to our social advances. Granting all the confusions and troubles that greet us with each issue of the newspapers, it remains true that man has made great progress in sorting out his human relationships. The cry "Who goes home?" which still adjourns the House of Commons, reminds us that not too long ago members required armed escort to protect them from the brigands who lurked between Westminster and the City. The constitutional experience of the American republic is impressive evidence that society does not always blunder. His Majesty's loyal opposition—the difference between political opposition and treason—is the basic treaty of political life in widening areas of the world. If science has made great contributions to man's well-being, the institution of contract has, in an unobtrusive way, made it possible. And it is deeply satisfying to recall that the daily lives of most people are saved from Hobbes' jungle by the presumption of good faith that infuses our relationships with one another.

To return to the four major successes, it seems interesting to note certain features that show how disparate they are. The first success—that of the physical sciences—is in a field where logic and quantitative measurement are dominant. The second—the dawning light of understanding of animate nature—is far less advanced, and it involves factors that are certainly nonquantitative and may well prove alogical. The third—the perfection of the codes of personal conduct—is curiously and unhappily more a matter of theory than of practice. I believe it was Chesterton who remarked that no one knows whether Christianity will work because no one has ever tried it. As an ex-mathematician, I would point out that one single clear exception proves that a presumptive general rule is incorrect, and I would therefore say that Chesterton's remark is characteristically vivid and interesting, but that it is false. The fourth success—man's enrichment of his life through the arts—presents features that are baffling to a scientist. Indeed, I am not sure that the word *success* really applies here, for success connotes a bad start and good progress. But the arts, as a previous AAAS president has pointed out, seem to constitute an almost completely non-accumulative part of experience. Rutherford had a great natural advantage over Faraday, and he over Gilbert; with respect to electric phenomena, both theory and the techniques of experimentation kept advancing, and each step was built on top of the preceding one. But Emily Dickinson had no advantage over Sappho. Each simply had words, the challenge of beauty, and the ineffable genius to condense, purify, and universalize experience.

Of these four major successes, I believe it is rather clear that the most tangible and obvious is the success of physical science. And this is an instance in which success and danger are close companions, as they often are. I do not refer here to the danger—ominous as it is—that science has unleashed forces that can physically destroy us. I refer to the more subtle danger that this success may mislead us concerning the real nature of science and its relationship to the rest of life and thus destroy something that is in the long run more important than a factory or a city, namely, our sense of value.

What made possible the great success that the physical sciences have experienced, particularly during the last century and a half? The explanation appears actually to be rather simple. *Physical nature*, first of all, seems to be on the whole very *loosely coupled*. That is to say, excellently workable approximations result from studying physical nature bit by bit, two or three variables at a time, and treating these bits as isolated. Furthermore, a large number of the broadly applicable laws are, to useful approximation, linear, if not directly in the relevant variables, then in nothing worse than their second time derivatives. And finally, a large fraction of physical phenomena (meteorology is sometimes an important exception) exhibit *stability*: perturbations tend to fade out, and great consequences do not result from very small causes.

These three extremely convenient characteristics of physical nature bring it about that vast ranges of phenomena can be satisfactorily handled by linear algebraic or differential equations, often involving only one or two dependent variables; they also make the handling *safe* in the sense that small errors are unlikely to propagate, go wild, and prove disastrous. Animate nature, on the other hand, presents highly complex and highly coupled systems—these are, in fact, dominant characteristics of what we call organisms. It takes a lot of variables to describe a man, or for that matter a virus; and you cannot often usefully study these variables two at a time. Animate nature also exhibits very confusing instabilities, as students of history, the stock market, or genetics are well aware.

If the successes of physical theory had remained limited to those highly useful but none the less essentially simple situations covered by two variable equations such as Ohm's law in electricity, or Hook's law for elastic deformation, or Boyle's law for volume and pressure of gases, or even to the vastly greater range of dynamic phenomena that are so superbly summarized in Newton's second law of motion, then it seems likely that mankind would have preserved a reasonable, take-it-or-leave-it attitude toward science. But two further things occurred.

Physical science pushed on to much more subtle and more complicated realms of phenomena, particularly in astrophysics and in atomic and then nuclear physics. And it kept on having successes. Second, physical science (and remember nowadays it is not really useful to discriminate between physics and chemistry) began to be applied more and more to certain limited sorts of problems of animate nature. Biochemistry, to take a very conspicuous example, began to deal successfully with phase after phase of the happenings within the individual cells of living creatures.

At the same time, of course, scientific theories kept getting more and more complicated and technical. Not only were they generally formidable to the public at large—scientific experts themselves had increasing difficulty in understanding anything outside their own specialities.

All this has tended to create a set of superstitions about science. These seem to be rather widely adopted by the public, and some of them even have adherents among scientists! These superstitions go something like this:

Science is all-powerful. It can just do anything. If you doubt this, just look around and see what it has done. A procedure known as "the scientific method" would in fact, if we only used it, solve all the problems of economics, sociology, political science, esthetics, philosophy, and religion. And the reason why science has been so successful, and the basis of confidence that it can go on to do anything whatsoever, is that science has somehow got the real low-down on nature and life. It has found out how to capture absolute truth, exact fact, incontrovertible evidence. Its statements are just "mathematically true," and in the face of that, you had better be confident and respectful, even if you are confused.

But science (to continue the superstitions) cannot be understood by ordinary folk. It is too technical, too abstruse, too special, and too different from ordinary thinking and ordinary experience. There is a special small priesthood of scientific practitioners; they know the secrets and they hold the power.

The scientific priests themselves are wonderful but strange creatures. They admittedly possess mysterious mental abilities; they are motivated by a strange and powerful code known as "the spirit of science," one feature of which seems to be that scientists consider that they deserve very special treatment by society.

Now these are dangerous misconceptions about science. If they were wholly untrue, if they were total and complete nonsense, then one could confidently await the general recognition of their fraudulent nature. But there is just enough apparent and illusive evidence in favor of these statements to give them an unfortunate vitality.

Let me list as briefly as I can a set of alternative statements which I believe to be more reasonable and accurate.

1) Science has impressively proved itself to be a powerful way of dealing with certain aspects of our experience. These are, in general, the logical and quantitative aspects, and the method works superbly for linear and stable physical problems in two or three variables. The physical universe seems to be put together in such a way that this scientific approach is exceedingly successful in producing a good, workable, initial description. And with that kind of solid start, physical science can then safely proceed to elaborate more sophisticated theories.

2) We simply do not yet know how far these methods, which have worked so well with physical nature, will be successful in the world of living things. The successes to date are very impressive. One feature after another that previously seemed to fall in a special "vital" category has usefully yielded to biochemical or biophysical attack. But it is also the case that we have as yet made only a beginning. How far the logical-quantitative method will succeed here, one would be rash to forecast, although the prospects do indeed seem extremely promising.

3) We have made small beginnings at extending the scientific method into the social sciences. Insofar as these fields can be dealt with in terms of measurable quantities, they seem to present closely intercoupled situations that can very seldom usefully be handled with two or three variables and that often required a whole hatful—for example, W. Leontief's input-output analysis of the U.S. economy deals with some 50 variables and regrets that it does not handle more. Science has, as yet, no really good way of coping with these multivariable but nonstatistical problems, although it is possible that ultrahigh-speed computers will inspire new sorts of mathematical procedures that will be successful in cases where the effects are too numerous to handle easily but not numerous enough or of suitable character to permit statistical treatment. If we try to avoid the many-variable aspect of the social sciences by using highly simplified models of few variables, then these models are often too artificial and oversimplified to be useful. The statistical approach, on the other hand, has recently exhibited—for example, in the stochastic models for learning—new potentialities in the field of human behavior.

4) It is, incidentally, not at all necessary that the particular analytic techniques of the physical sciences be forced upon biological or social problems with the arrogant assumption that they can and should make unnecessary other types of insight and experience. During the recent war, an extremely useful collaboration was developed, known often as operations analysis, in which reasoning of a mathematical type was applied to certain aspects of very complicated situations, but with no expectation that judgment, experience, intuition, or a vague sort of general wisdom would be displaced or superseded—rather only that these would be aided by whatever partial light could be furnished by quantitative analysis.

5) An important characteristic of science, which we must note in passing, is its incapacity to be impractical. The most far-reaching discoveries and the most widespread useful applications flow regularly out of ideas that initially seem abstract and even esoteric. These ideas arise out of the unguided and free activity of men who are motivated by curiosity or who, even more generally, are thinking about scientific problems simply because they like to. The way in which apparently aimless curiosity stubbornly refuses to be foolish and leads to important goals doubtless seems strange or even incredible to some persons. The eventual usefulness of the initially impractical is widely held to be a very special feature of science, but I am not so sure of this. I think that apparent impracticality is more generally important than we are inclined to suppose.

6) Science presents the kind of challenge that attracts to it young men and women who tend to have a rather high degree of a certain kind of intelligence. Since this particular kind of intelligence is relatively easy to recognize and measure, and since many other types are subtle and illusive even though perhaps more important, we tend to adopt this one type as the norm. In addition, this particular type of intelligence leads rather promptly to tangible results. These circumstances lead to the con-

clusion, which is then something of a tautology, that scientists are more intelligent than other people. This may or may not be true; more important, however, it may be neither true nor untrue in the sense that the attempted comparison is meaningless.

7) However, despite their appearing to be so bright, scientists are not special creatures: they are people. Like lots of other people, they are good at their own tasks. Off their jobs they seem, as Shylock remarked in another connection, "to be fed with the same food, hurt with the same weapons, subject to the same diseases, healed by the same means, warmed and cooled by the same winter and summer" as other men are. When you prick them they do indeed bleed.

A. V. Hill, while he was president of the British Association for the Advancement of Science, stated: "Most scientists are quite ordinary folk, with ordinary human virtues, weaknesses, and emotions. A few of the most eminent ones indeed are people of superlative general ability, who could have done many things well; a few are freaks, with a freakish capacity and intuition in their special fields, but an extreme naiveté in general affairs. . . . The great majority of scientists are between these groups, with much the same distribution of moral and intellectual characteristics as other educated people."

8) One rather accidental fact has led many to think that scientists are strange and special, and this is the fact that scientists often use a strange and special language. Science does find it desirable to use very many technical words, and it has indeed developed, as a matter of saving time, a sort of language of its own. This gives to science an external appearance of incomprehensibility that is very unfortunate. The public need not think itself stupid for failing intuitively to grasp all this technicality. Indeed, what has developed is not so much a language as a series of very specialized dialects, each really understood only by its inventors. "On faithful rings" is not a sociological discussion of marriage but an article on modern algebra. The "Two-body problem for triton" is not mythology, but physics; a "folded tree" is not a botanical accident, but a term in telephone switching theory.

9) If scientists are human, so also is science itself. For example, science does not deserve the reputation it has so widely gained of being based on absolute fact (whatever that is supposed to mean), of being wholly objective, of being infinitely precise, of being unchangeably permanent, of being philosophically inescapable and unchallengeable. There seem still to be persons who think that science deals with certainty, whereas it is the case, of course, that it deals with probabilities. There seem still to be persons who think that science is the one activity that deals with truth, whereas it is the case, of course, that—to take a very simple example— "the true length of a rod" is so clearly not obtainable by any scientific procedure that, insofar as science is concerned, this "true length" remains a pleasant fiction.

I could document this particular point at length, but will restrict myself to three quotations from the relatively mature fields of physics, astronomy, and mathematics.

Edmund Whittaker said of theoretical physics: ". . . it is built around conceptions; and the progress of the subject consists very largely in replacing these conceptions by other conceptions, which transcend or even contradict them."

Herbert Dingle, in his retiring address as president of the Royal Astronomical Society, said: "The universe . . . is a hypothetical entity of which what we observe is an almost negligible part. . . . In cosmology we are again, like the philosophers of the Middle Ages, facing a world almost entirely unknown."

Alfred North Whitehead has stated: "While mathematics is a convenience in relating certain types of order to our comprehension, it does not . . . give us any account of their activity. . . . When I was a young man . . . I was taught science and mathematics by brilliant men . . . since the turn of the century I have lived to see every one of the basic assumptions of both set aside."

10) These quotations indicate that the ablest scientists themselves realize the postulational and provisional character of science. Perhaps not so widely recognized or accepted is the extent to which the development of Western science, rather than constituting a uniquely inevitable pattern, has been influenced by the general nature of Greco-Judaic culture, including especially the standards, arising within that tradition, of what is interesting and important.

Confronted by the totality of experience, men select the features that seem interesting and important—and the criteria for interest and importance arise not just or even primarily within scientific thought, but rather within the entire cultural complex. One then seeks to find a way of ordering this selected experience so that the end result is acclaimed as satisfying and useful—again as judged within the total culture. This process has different possible beginnings and different possible procedures; so, of course, it has different possible end results. Clyde Kluckhohn has remarked, "What people perceive, and how they conceptualize their perceptions is overwhelmingly influenced by culture." H. M. Tomlinson said, "We see things not as they are, but as we are."

If, for example, a culture almost wholly disregards physical suffering, considers the present life an unimportant episode, and places a very high premium on prolonged mystic contemplation, then this viewpoint regarding values does more than, for example, underemphasize modern scientific medicine (using all these words in the Western sense). It produced something that is different *in kind;* I know of no criteria that justify calling one kind good and intelligent, and the other poor and ignorant.

Chang Tung-San, a Chinese philosopher, has said: "Take Aristotelian logic, for example, which is evidently based on Greek grammar. The differences between Latin, French, English, and German grammatical form do not result in any difference between Aristotelian logic and their respective rules of reasoning, because they belong to the same Indo-European linguistic family. Should this logic be applied to Chinese thought, however, it will prove inappropriate. This fact shows that Aristotelian logic is based on the Western system of language. Therefore we should not follow Western logicians in taking for granted that their logic is the universal rule of human reasoning."

If this general line of thought seems to you either interesting or improbable, I urge you to read some of the fascinating papers of Benjamin Lee Whorf and of Dorothy D. Lee on the value systems and the conceptual implications of the languages of various American Indian tribes. Whorf, for example, points out that the Hopi Indian language "is seen to contain no words, grammatical forms, constructions or expressions that refer directly to what we call *time*, or to past, present, or future, or to enduring or lasting, or to motion as kinematic rather than dynamic. . . . At the same time the Hopi language is capable of accounting for and describing correctly, in a pragmatic or operational sense, all observable phenomena of the universe."

11) The ten preceding numbered comments concerning certain general characteristics of science all contribute, I believe, to a major conclusion—that science is a very human enterprise, colored by our general ideas, changeable as any human activity must be, various in its possible forms, and a common part of the lives of all men.

Indeed, even the impressive methods that science has developed—methods which sometimes seem so formidable—are in no sense superhuman. They involve only improvement—great, to be sure—of procedures of observation and analysis that the human race has always used. In the appeal to evidence, science has taught us a great deal about objectivity and relevance, but, again, this is refinement of procedure, not invention of wholly new procedure.

In short, every man is to some degree a scientist. It is misleading that a tiny fraction of the population is composed of individuals who possess a high degree of scientific skill, while most of the rest are indifferent or poor scientists. This creates the false impression that there is a difference in kind, when it is actually only one of degree.

If, when a window sticks, you pound it unreasonably, or jerk so hard that you hurt your back, or just give up in ignorant disgust, then you are being a poor scientist. If you look the situation over carefully to see what is really the matter—paint on the outside that needs cutting through, or a crooked position in the frame—then you are being a good scientist.

Even primitive men were scientists, and in certain aspects of accurate and subtle observation and deduction it would probably be hard to beat the ancient skilled hunter.

Indeed, one important contrast between the savage and the professor is simply that modern scientific methods make it possible to crystallize our experience rapidly and reliably, whereas primitive science does this clumsily,

133

slowly, and with much attendant error. But it is, after all, well to remember that ephedrine is the active principle in an herb, Ma Huang, that has been empirically employed by native Chinese physicians for some 5000 years. Certain African savages when they moved their villages did take with them to the new location some dirt from the floor of the old hut. Moreover, it is true that they said that they did this to avoid the anger of their gods who might not wish them to move, fooling them by continuing to live on some of the same ground. But the fact remains that by this process they brought to the new location the soil microorganisms that continued to give some degree of protection from certain ailments. We quite properly honor Fleming and Florey, but Johannes de Sancto Paulo, a medical writer of the 12th century, did prescribe moldy bread for an inflamed abscess. "We are all scientists," Thomas Huxley said, because "the method of scientific investigation is nothing but the expression of the necessary mode of working of the human mind."

Let us now back away from the trees and look at the forest. Where have we arrived in this discussion?

I have just listed 11 points that, in my judgment at least, fairly characterize science as a universal human activity. These comments do not support the concept of science as some sort of super creed, magical and mysterious as it is all-powerful, arrogant from its successes, and avid to invade and conquer, one after another, all the fields of human activity and thought. This viewpoint does not justify the notion that science is so special as to be unique, as well as so curious as to be incomprehensible. This does not depict scientists as strange creatures who are in one sense so objective, judicial, and precise as to be incredible, and in another sense so apart from life as to be selfish and sinister. This does not set up quantitative analytic Western science as the only valid way in which man may approach and interpret experience.

On the contrary, these descriptive comments picture science as the servant of man, not his master; and as a friendly companion of art and of moral philosophy. This is a science that is the way it is because man wants it to be that way. It is a natural expression of both his curiosity and his faith.

If the public could be brought to understand and appreciate this position concerning science and scientists, I do not think that so many persons would harm this great enterprise of ours with a combination of mistrust, fear, and overestimation. I do not think that so many would treat scientists one-third of the time as amusing but beneficial eccentrics, one-third of the time as sorcerers, and one-third of the time as irresponsible rascals. I do not think that so many would view scientists as careless dabblers with danger or as a selfish minority that, to quote a nationally syndicated columnist, "hold they are an extra special group not tied down by the obligations and rules under which the rest of us work. Hundreds of them are now bellyaching about the Oppenheimer verdict and saying it ruins their morale and makes them hard to get. What goes with those birds?"

Or consider another newspaper writer who opened one of his columns with the sentence, "We Americans have been confronted with an arrogant proposition that persons presuming to call themselves intellectuals, and particularly those who claim the title of scientist, are a superior cult entitled to deference or even homage from the common man." One of our greatest universities takes a sound and courageous stand, and a newspaper writer complains, "Harvard has a peculiar fondness nowadays for putting security and the safety of the nation second to their fancy ideas of importance." If some speak out against the climate of fear resulting from the stupidities and iniquities of what is misnamed as the security system—doubly misnamed since it is not a system and does not achieve security—then their protest is labeled, as it was by Eugene Lyons in the *Saturday Evening Post*, as "the mock-heroic posture of this close-knit band of Cassandras"; he insultingly adds that these protesters do not themselves seem to have suffered, for "not one of them has as yet been muzzled, lynched, or denied his due royalties."

Anti-intellectual views such as these are widely expressed in those newspapers that combine a wide circulation with a narrow intellectual viewpoint, in some very popular national magazines, and even, one reports with shame, by highly placed persons in Washington.

It is hardly necessary to argue, these days, that science is essential to the public. It is becoming equally true, as the support of science moves more and more to state and national sources, that the public is essential to science. The lack of general comprehension of science is thus dangerous both to science and to the public, these being interlocked aspects of the common danger that scientists will not be given the freedom, the understanding, and the support that are necessary for vigorous and imaginative development. It is, moreover, of equally grave importance that science understand itself.

There are persons who are pessimistic concerning the prospects of materially improving the public understanding of science, and even the understanding that one branch of science has of the other branches. If one subscribes to the falsities and exaggerations that I stated in the first part of this article, then he could properly be pessimistic. If, on the other hand, he accepts the broader, more liberal, more human and humane view that I have advanced here, then—or at least so it seems to me—he can be very optimistic.

When David Brewster, a century and a quarter ago, was one of the prime movers in founding the British Association for the Advancement of Science, he said, "The principal objects of the Society would be to make the cultivators of science acquainted with each other, to stimulate one another to new exertions—to bring the objects of science more before the public eye and to take measures for advancing its interests and accelerating its progress."

This is a challenge which our own Association has always sought to meet. It is a challenge which, at this moment in history, requires renewed zeal and ever-renewed patience. Speaking of the present-day scientist, J. Bronowski has said, "Outside his laboratory, his task is to educate us in what goes on inside it, and to give it a meaning for us. In a world in which statesmen as much as voters are ignorant of the simplest implications in science, this is a formidable responsibility . . . [the scientist] has no other choice today but patiently to become a teacher, in a world in which distrust and prejudice are free. . . . There is no alternative to an informed public opinion: and that can exist only where scientists speak to voters and voters accept their responsibility, which is to listen, to weigh, and then to make their own choice."

If, as I believe, the sciences and the arts are lively and noncompetitive partners in the business of life, it is appropriate that we close, not with a scientist, but with a great artist. "Our privacy," Faulkner says, "has been slowly and steadily and increasingly invaded until now our very dream of civilization is in danger. Who will save us but the scientist and the humanitarian. Yes, the humanitarian in science, and the scientist in the humanity of man."

The Inexorable Problem of Space

Paul B. Sears
Indianapolis, 1957

Paul B. Sears (1891–   ), *botanist and ecologist, received his Ph.D. from the University of Chicago in 1922. He taught botany at a number of colleges and universities, including the University of Oklahoma and Oberlin College. Sears has done research in a variety of areas during his career, including climatic sequences, fossil pollen, cultural patterns in ecology, and problems relating to conservation. From 1950 until his retirement in 1960, Sears was head of the new graduate research program in the conservation of natural resources at Yale University. He has been described as a biological statesman who sees the life sciences against the background of human affairs, a viewpoint reflected in both his research work and his nontechnical writings. Sears is the author of many books, including* Deserts on the March *(1935);* This Useful World *(1941); and* The Biology of the Living Landscape *(1964). He was president of the AAAS in 1956.*

Discovery and communication are the two prime obligations of the scientist. On occasions such as this, however, the scientist has the added opportunity to examine broad issues in the light of his peculiar knowledge and experience. This I propose to do with respect to that limited segment of space in which we live, move, and have our being. For my subject was chosen long before man's most recent and dramatic invasion of outer space.

My thesis is that, among the practical problems of humanity today, our relation to immediate space is of critical importance. In developing this idea, I shall try to show that our applications of science have been both restricted and short-sighted. In terms of moral choice, we have looked upon science as an expedient rather than as a source of enlightenment.

To be specific, our very proper concern with the applications of mathematics, physics, and chemistry may be clouding the fact that we need biology in general and ecology in particular to illuminate man's relation to his environment. At present the biological sciences are largely sustained as utilities in medicine and agriculture, the social sciences for dealing with immediate ills. But we must not forget that all science is needed to guide the process of future evolution—cultural and physical—now so largely in our own hands. The nest of anti-intellectualism is being warmed by the ignorant, but some of the eggs in it may have been placed there by those who should know better.

I have no quarrel with the exploration of outer space. It is a legitimate and challenging subject for scientific inquiry and bold experiment. Our optical and mathematical studies of it have long since given us that basic confidence in order without which there could be no science. But, as we extend our astronomy by whatever celestial acrobatics we can get away with, I should like to see some consideration given to relative values. We have a vast amount of unfinished business at our feet. The golden moment for the pickpocket comes when everyone at the county fair is craning his neck at the balloon ascension.

So far as the skies are concerned, we are feeling the natural soreness that comes from losing a sporting event we thought was in the bag. Actually, if my information is correct, the Russians had explained that they intended to launch a satellite, had indicated its probable size, and have promised to share the knowledge so gained. Since any ray of light should be welcomed in an atmosphere of gloom, it may help to recall that our Olympic athletes, in the face of leading questions from their interviewers, had nothing but respect to offer for the conduct of their Russian rivals.

Of course our present concern is much more than simple chagrin at losing a contest. What has happened in outer space raises a question about how outer space will be allocated and controlled. We fear, not unreasonably, that whoever controls the space around the earth can impose his will upon all who live on the earth's surface.

Though we grant freely the military significance of space experiments, our present hysteria seems to me to indicate an even deeper source of insecurity. We are beginning to sense that the elaborate technology to which we are so thoroughly committed makes us peculiarly vulnerable. And we are not wholly confident that the ideals of our civilization—so reasonable to us—will really stand up to free competition with other systems of thought. To the extent that this is true we suffer from an initial handicap of morale.

The pattern of conflict is much the same, regardless of scale. Whether one is watching small boys in the school yard or great powers in the world arena, the preliminaries are marked by bad manners and vituperation on both sides. Missiles are piled up and seconds are assembled, the advantage going to the cooler, less hysterical side. The contestant who gets rattled is asking for trouble.

I do not envy our public servants charged with the delicate business of managing international relations. But I am firmly convinced that unless one is determined on war, there is merit in self-restraint and good manners, as well as in prudent measures of self-protection. I am also convinced that the choice of policy is not limited to boasting and belligerence on the one hand or craven appeasement on the other. We have no monopoly on self-respect and other human virtues, nor is there any merit in debasing the original meaning of the word *compromise* as we have done. We should deplore every display, whether by statesman or journalist, of dunghill courage that lessens the hope of mutual understanding, good-will, and ultimate collaboration among human beings.

So far as purely domestic problems go, our almost hypnotic concern with outer space comes at a bad time. Outer space is one more item that diverts attention and energy from the prosaic business of setting our terrestrial space in order. And it has fostered an incredible type of escapism that must be experienced to be believed. One hears too frequently for comfort the sober assertion that we need not worry about depletion of natural resources, now that interplanetary travel is just around the corner! If such a comment came from jesters or cranks, it could be disregarded. But we hear it uttered with the solemnity and assurance of the true believer. No doubt we shall continue to hear it, despite the chilling analysis by Arthur Clarke, the British astronomer, in the November 1957 issue of *Harper's* magazine.

Actually this obsession is not a detached phenomenon. Rather it is the culmination of a new faith—the belief that technology will solve any problems that may confront humanity. Curiously, it comes at a time when the scientist is more suspect than he has been since the days of witchcraft and alchemy, as recent opinion studies show. A high proportion of people consider scientists to be queer fish, if not inhuman and immoral. For a parallel we would have to think of a religion which wants the favor of its gods but does not trust them for a moment.

I do not question the tremendous accomplishments and future possibilities of technology. I yield to no one in my admiration for the cleverness, manual and intellectual, of those who apply science to meet the needs of mankind. But faith in technology is not faith in science or sympathy with the creative impulse of the scientist. The direction in which science is applied depends upon the values of the culture applying it even while science is in turn modifying the culture.

Our present applications of science are selective and opportunistic, neither wholehearted nor balanced. We are applying it out of all proportion to the elaboration of consumer goods, often to such an extent that vast sums must go into persuading people to desire what they have not instinctively wanted. The making of things has become so facile that their sale creates major problems in advertising and credit. As Max Beerbohm once put it, "Buy advertised goods and help pay the cost of advertising." Some of the keenest satire on advertising has come from advertising men themselves. Raymond Loewy, the famous automobile designer, has protested the corruption in car design that has resulted from too great facility— traditionally the death of any sound art. The current models waste space, materials, and fuel, violate good taste, and impose needless economic burdens on the public. And while we are applying science in this manner, we are blandly ignoring its highest function, which is to give us perspective and to inform us about what we are doing to ourselves. Even the scientist, as Kubie has so ably shown, suffers from his lack of self-analysis.

On the whole, a man's actions are a response to his idea of the kind of world he thinks he is living in and to his concept of his own nature. That this is true is shown very practically in the history of human thought: No great religion is content merely to lay down maxims of conduct; it also develops its own cosmogony, its own pattern of the universe, to justify those maxims.

Our present attitude toward terrestrial space exemplifies with peculiar clarity our selective use of science. For living space, if we consider both its extent and quality, subsumes all other resources, being in that respect equivalent to the economist's technical concept of land. Yet the power of applied science has been overwhelmingly employed to exploit space, while those aspects of science which could illuminate its wise and lasting use are still largely ignored.

I am assuming at the outset that the human adventure on this planet is worth our best efforts to keep it going as long as possible. I am also assuming that man is capable of responsible judgment and conduct and that he has at hand much of the important basic information he needs. Finally, I am assuming that it is not enough for man to live by bread alone but that intangible, as well as tangible, values are necessary to justify his persistence. If this be true, the question is, not how many people can exist on earth, but what kind of a life will be possible for those who do.

From New Jersey to Oregon one sees great egg factories, where highly selected strains of poultry are confined at maximum density and with maximum efficiency. Every need—nutritive, environmental, and psychological—is taken care of. These gentle, stupid birds have no responsibility but to stay alive and do their stuff. Yet they are at the mercy of any break in an elaborate technological mesh that keeps them going. And should a stranger burst abruptly into their quarters, the ensuing panic would pile them up in smothering heaps in the far corners of their ultramodern apartment. The underprivileged, pretechnological hen ran many hazards, but at least she had the freedom to scratch around for food and a sporting chance to dodge under a bush to evade the swooping hawk.

People, of course, are not poultry, but they are living organisms, subject to the limitations inherent in that condition. I am unmoved by any protest against applying biological analogies to human society. Analogy is one of the most powerful tools of the scientist. From physics to physiology, and notably the latter, analogies suggest our models which we must then test and either accept or reject as the evidence may dictate. And besides, man is a living organism, as I have said.

Fortunately, in considering man's relation to terrestrial space, our models do not all come from observing other forms of life. We have some impressive ones furnished by our own species. Let us reserve them, however, for the present and look at the other living things. Here from students of bacteria, trees, insects or any of the sundry groups of vertebrate animals—fish, fowl, or mammal—we get the same story. No known form of life has been observed to multiply indefinitely without bumping up against the limitations imposed by the space it occupies. These limitations involve not only quantity but quality. And quality rests upon the pattern of that complex of factors, whether known or unknown, that are necessary to sustain the species in question. So far as environment is concerned, an ancient bit of wisdom sums up the situation: "A chain is no stronger than its weakest link."

This principle was recognized by Liebig in his famous law of the minimum: the growth of a crop is determined by the essential nutrient available in least quantity. It was restated more precisely by Blackman in his law of limiting factors: physiological processes are limited by the least favorable factor in the system of essential conditions. These statements rest upon controlled experiment. They are independent of the circumstance that an English political economist and parson, Malthus by name, had suggested that human populations did not, in fact, increase indefinitely beyond certain limitations of environment.

It should be emphasized, however, that the writings of Malthus did give the necessary clue that enabled Charles Darwin to formulate a reasonable explanation of the mechanism involved in the origin of the species. Since there remain many misconceptions with regard to both Malthus and Darwin, it may be well to review the thesis of the latter. This thesis has never been successfully controverted, although there are perennial headlines "Darwin Refuted" whenever some detail of his work is brought into question. Briefly, Darwin had noted the universal tendency of organisms to vary and to transmit these variations to their offspring. Our knowledge with respect to these matters is now being applied daily by plant and animal breeders with the same effectiveness with which the phase rule is used in chemical engineering.

Darwin's second point was that organisms tend to reproduce far beyond their capacity to survive. This again in fact occurs, and is a matter of household knowledge among those who, as scientists, observe living organisms. The tagging of fish that return to the place where they began migration reveals that, despite the thousands of eggs laid by each female, not more than a few adults from each batch survive to make the return journey.

Anyone who has observed, year after year, the nesting of robins in his yard has noted the consistent toll—from cats, jays, crows, and accidents—that serves to keep numbers down. And though the clutch of eggs is fairly uniform from year to year—implying a potential doubling of the robin population—the number of nests does not increase significantly, nor do these birds spread beyond a well-defined territorial range. Even though predators might fail to control their numbers, competition within the species would establish a threshold of limitation, as it did for deer when wolves were eliminated.

The final point made by Darwin was that the relatively small proportion of individuals surviving did so, not merely by random chance, but largely because they were those best fitted to cope with their environment.

The less favorably endowed tended to be eliminated. Thus the better adapted lived to transmit their favorable variations. In this way he accounted for two great riddles of living nature—the immense variety of living things and the remarkable adjustments they show.

It is not my object here to justify Darwinian theory. It is enough to say that the theory coordinates more information than any alternative that has been proposed. This is all we have a right to ask in science. We need emphasize only one corollary—that the pattern of environment is built solidly into that of life. Survival rests upon this relationship. No organism known to biologists has ever, so to speak, had things completely its own way. Some, of course, are more "successful" than others, as the late L. O. Howard indicated in his famous dictum that the last survivor on earth would be a living insect on a dead weed.

The rub comes when we attempt to extend these principles to our own species. As life has advanced from simple beginnings, it has played an increasing role in geological processes. Man is no exception. He is a world-wide dominant, the first such species in earth history. And through advancing technology he is producing tremendous changes. That this should confer a sense of power is understandable. But power is not the same thing as control. Only when power is balanced by responsibility is there control, as the record of our highway accidents attests. The corrupting effect of irresponsible power is an axiom in human history.

Is there any reason to believe that man is exempt from the rules that apply to living organisms in general? Or does the difference lie in his ability to learn those rules and profit by respecting them? Can we make use of known physical and biological principles in discussing problems that involve man? Can we view psychological and sociological considerations in the light of simpler and more obvious ones, or must we rely solely upon a higher level of discourse when we speak of man? These are not idle questions. I have mentioned the indignant protests against applying "biological analogies" to sociological problems. But it is one thing to hold that man is *merely* a physicochemical system, or *merely* an animal, and quite another to insist, as I must, that he is a physical phenomenon, and a biological one too, whatever else he may be.

Man's physical body occupies space, somewhere between two and four cubic feet of it. At his present rate of increase in the United States, he is set to double the aggregate volume occupied by human bodies in about 41 years. Continuing at this rate, it would be less than 700 years— say 22 generations—until there is standing room only, with each space 3 by 2 feet, or 6 square feet, occupied. On this basis there is room for exactly 4,646,400 people in each square mile. I have perhaps been over-generous in estimating the per capita area, but I did wish to leave space enough to permit each individual to reach in his pocket for rent money when it falls due. A little after this the hypothetical human population would weigh more than the planet.

In thus giving rein to imagination I have in mind sundry pronouncements regarding the potential capacity of the earth, some of them to the effect that by proper scientific management it can take care of any conceivable increase in population. The numbers I have mentioned are both conceivable and begettable. The question is, are they supportable?

The most densely populated continent is Europe, with 142 people per square mile, as against Asia with 78, although the most densely populated areas are as yet on that continent. Australia follows with 31, then North America, including great areas of desert and tundra, with 23, while Africa and South America are nearly tied, with 17 and 16, respectively. The figure for the United States is 51, intermediate between that of Australia and Asia. Evidently cold fact, as so often happens, has not kept pace with theory. Either people do not breed as fast as they might, or survival rates are not what they could be. Actually both of these things happen, and in curious combinations. We may, I think, allow the battered bones of the Reverend Malthus to rest in peace as we examine a few case histories quite briefly.

First, however, let us retrace our steps for a glimpse at what we pleasantly call the lower orders of life. Abstracting an item from the valuable studies of Thomas Park, we learn that when populations of flour beetles reach certain densities, their rates of increase drop sharply. Among other things, these animals begin to eat their own eggs and pupae, a very effective way of slowing down the operation of the compound interest law. Whether this practice is due to a craving for food and water or simply to the fact that hungry beetles bump into eggs oftener than before, we do not know.

The lemmings in Alaska are likewise instructive. These small rodents, living and breeding under the snow, have a kind of pulsating population record, abundance alternating with low density in fairly regular fashion. With summer melting, they are preyed upon by a variety of animals, including the Arctic fox and snowy owl. A third predator, the jaeger, a kind of sea hawk somewhat resembling a gull, has been studied by Frank Pitelka, who reported on it at the Berkeley meeting of the AAAS. When the lemming population is low or average, the jaegers space their nests and consume their prey in an orderly manner. But when the lemmings are at a peak, so that food should be no problem, the jaegers spend so much energy quarreling over nesting space and food that relatively few of them raise normal broods. So their numbers decline, but not primarily from lack of food. They do not urbanize well—or shall we say that when they attempt to urbanize they pay the usual penalty of a greatly lowered reproductive efficiency? For it is, I believe, an open secret that few cities of major size have heretofore maintained their population by their own birth rate—a situation that is probably changing through the rapid development of suburban life. Perhaps it is time for some modern Aesop to instruct us on manners and morals, using for that purpose the verified behavior of animals instead of their imagined words. Certainly we learn that for the jaegers plenty is not an inevitable road to biological success.

Yet the idea of plenty—in food in particular, in energy and minerals to a lesser degree—dominates the discussion by scientists of man's future. Some of this material is excellent, notably that by Harrison Brown, who not only understands the physical sciences but has biological sense and a conscience to boot. Too few, however, bother to read the fine print and observe the *if's* in such analyses as his. Those who, like Osborn, Cook, Sax, and Vogt, concern themselves with space and numbers are written off as "pessimists," as though the fixing of a label adjudicates the issue and solves the problem.

It is the merit of the men named, including Brown, that they have raised not only a material but a moral issue that is too often neglected by those who proudly label themselves "optimists." The question is not only how much but what kind of life will be possible if humanity continues to hurtle along its present course. Russell, the Huxleys, and Berrill have all warned us of the inevitable loss of freedom and personal dignity that must follow the multiplication of numbers and the depletion of resources.

The findings of archeology are in agreement with recorded observations of prefarming cultures about the space requirements of hunters, fishers, and gatherers. For such folk the space requirements are great, by modern standards, being no less than three to five square miles per person where conditions are most favorable. The best estimates for pre-Columbian United States, even with such agriculture as it possessed, do not reach three million in about the same number of square miles. Specifically, the state of Ohio, some 40,000 square miles, mostly fertile and well-watered, does not appear to have supported more than about 15,000 Indians at the time of European discovery. Even the Basin of Mexico, with a highly efficient system of horticulture and an imposing array of domesticated plants, did not have numbers exceeding a million—one-third the population of the present Mexico City, which occupies only a fraction of the modern basin.

Yet we know that this rather moderately concentrated population experienced pressures of various kinds during the centuries preceding 1519. However the situation might be rationalized, the limitations of space, with regard to both extent and quality, were stern and tangible within the Basin of Mexico. The ancient chronicles are a record of floods, drouths, volcanism, and hunger. Toward the end of the Aztec Empire, in a desperate attempt to placate the angry gods, human sacrifice was stepped up until it reached scores of thousands—suggesting the rate of emigration that today serves to stabilize the population of Ireland, whose chief export is people.

Our judgment of the whole history of agriculture has been revamped since the 1930's. Dale and Carter have done this brilliantly, showing that every great center of power and civilization has been based squarely upon fertile space, and tracing the parallel decline of culture and the nutrient capacity of the soil.

Certainly human communities have, as a matter of record, more than once run hard into the physical limitations of their environment. Often they have intensified these limitations by their own activities. That man can preserve and even enhance the potential of his environment, I do not question. But I see no warrant for asserting that he has often done it or can do so indefinitely under his present pattern of behavior.

Limiting factors are not necessarily physical in the strict sense. Cultural disruption and spiritual discouragement may likewise act as restraints. This is believed to explain the well-known decrease of the native Indian population during the century following the Spanish conquest of Mexico. With little to live for, people may simply not have families, whatever the physiological facts and urges may be. Another instance is that of the slave population in Jamaica prior to 1842. The white population, numbering less than one-tenth that of the slaves, vigorously discouraged breeding among the slaves, since it was cheaper to buy new slaves than to propagate them. Nevertheless, the apathy toward life, attested by the high suicide rate among victims of the slave trade, is believed by competent authority to have been an important factor in the low effective reproduction rate among these pathetic humans.

If we come closer home, we have the significant drop in family size during our own depression of the 1930's. In this instance, the slow-down cannot be attributed to pressure from the physical environment, for the depression preceded the great drouth. Even then there was no real scarcity of food, merely a breakdown in the mechanism for its economic distribution. Presumably the direct pressure came from cultural anxiety, or what is sometimes called "social shock." Even the "recession" of 1949–50 produced a measurable effect, total births in 1950 numbering 17,000 less than in 1949.

We have, too, the earlier decrease in the British birthrate about 1921. This was the year in which Marie Stopes, already famous as a paleobotanist, enlightened the public on responsible parenthood. It was also a time of high post-war prices, and subsequently a time of flaming individualism. But it was not, so far as I know, a period of physical pressure from environmental forces. Have myself reared a family during the 20's and 30's, I can testify that in our own country there were many cultural pressures, neither physical nor economic, that encouraged one-child or at most two-child families. Not least among these pressures was increasing focus upon the personality and development of the individual child, at times to the point of morbid sentimentality.

Cultural influences can also act in the opposite direction, the classical instance being in the scriptural injunction to be fruiful and multiply and replenish the earth. Today, despite the staggering cost of education and the increasing cost of food—unchecked by our continuing agricultural surplus—the four-child family is in vogue. Incredible though it may sound, it is through the influence of fashion (call it example or prevailing custom if you prefer) that many modern families work out their response to the problem of population and space.

Certainly the record suggests that population density is influenced both by the physical and the cultural environment. However these may operate, either singly or in conjunction, they find expression in the behavior of individuals, and individuals differ greatly. Indeed, one of the most difficult of problems is to sort out the strands—cultural, physiological, intuitive, and rational—that are interwoven into the fabric of individual values and conduct. As Russell has pointed out, and as those who style themselves "human engineers" know only too well, the new psychology has little comfort to offer about the importance of reason in human conduct. This would be especially true among those least capable of using it, yet I, for one, would not give up what confidence we have in it.

Coming now more specifically to the problem of space, we find that the grim facts in certain countries which we euphemistically call "under-developed" speak for themselves, as anyone who has visited the Orient, the West Indies, or certain portions of Latin America must honestly admit. Humane and successful efforts to improve health conditions in such areas have, to date, merely intensified the problem, while equally high-minded efforts to improve food production and distribution have only deferred a solution. Ceylon, where disease control has resulted in doubling the population in less than a score of years, is a classical example. Meanwhile, food production has not kept pace, and the usable area of the island has been increased only very slightly through drainage of malarial swamps.

Perhaps the one bright spot in this gloomy picture is that many of the leaders in these crowded countries are now frankly recognizing the problem and trying, according to their various lights, to face it. But while I would not suggest for a moment that we allow them to stew in their own juice, I do suggest that our own problem deserves more attention than it is getting. The very fact that we have a margin of safety not enjoyed in many parts of the world is both a challenge and an opportunity. Let me recite a few facts, even though they may be familiar.

That the productivity of our agriculture can be increased far beyond the limits of the present surplus is not questioned. But each increment in production calls for increasing capital outlays. The investment in machinery, to say nothing of that in fertilizers, feed supplements, maintenance, taxes, and insurance, frequently approaches the value of the land. The knowledge, skill, and competence of the successful farmer today rivals that tolerated in the practice of medicine fifty years ago. In that interval our farm population has diminished by more than a half, being now less than 20 percent of our total population. The pressure to keep costly machinery earning its way often results in extensive operations at the cost of personal attention to those details which prevent deterioration of the whole enterprise, and which, in the end, may make the difference between profit and loss. So meager is the margin that a significant and growing number of model farms are now owned by industrialists and other people of means to permit legitimate losses on their tax returns. So far as our ultimate food and fiber supply is concerned, we need not expect something for nothing.

The late Robert Salter, surely a very conservative individual, pointed out that the high yields from hybrid corn were definitely being obtained at the expense of soil fertility. In the corn belt, yields of 100 bushels per acre are now about one-third as frequent as they once were. My guess is that farm surpluses will be only a memory within two decades.

Alternative methods of production are, of course, being proposed and investigated. Most of these involve increasing dependence upon elaborate technological devices, hence increasing energy, capital, and maintenance costs. Equally serious is the increasing vulnerability that comes from utter dependence upon elaborate technological systems. This can be illustrated by what has occurred when a brave and competent army, trained to rely solely upon mechanical transport, has faced in difficult terrain an enemy hardened to simpler and more primitive methods. It was illustrated by the comparative ease with which the Ozark hill people adjusted to a depression while their highly dependent urban neighbors were thrown completely out of gear for a long period of time.

I forbear to recite what would happen to some of our great urban centers in the event of certain entirely possible technological failures. This is information which ought to be classified if it could be. In October I observed the confusion following a two-hour power failure in the Grand Central area in New York City. Four days later an accident to a single car on the Merritt Parkway in Connecticut delayed traffic for an hour, during which time seven miles of motor cars were halted bumper to bumper. The analogy between extreme urbanites and the denizens of the egg factories mentioned earlier is too close to be comfortable. No doubt the subconscious realization of this accounts to some degree for the difference between our present mood and that of the Turks and Finns. These sturdy people proceed phlegmatically about their simple way of life in spite of their hazardous geographical position.

We too are a brave and peace-loving people. It is entirely possible that we are not so much moved by fear of an enemy as by lack of confidence in the structure of a system in which we are so deeply committed and involved. What I am saying is inspired by those who see in technology the complete answer to the world's problems. For I do not doubt that technology, like a human being, has the defects inherent in its own virtues. If, as I believe, it should be our servant and not our master, its advancement should be in the light of all scientific knowledge and not merely of those facets which are of immediate use. The biologist who attempted to apply his knowledge in defiance of known physical principles would be laughed out of court. Yet we seem singularly trustful of engineering projects carried out in disregard of ecological principles.

The most obvious and acute pressure upon space is in our great cities and surrounding metropolitan areas, whose existence and extension depend upon technology. They and the associated industries and highways that connect them are absorbing agricultural land in the United States at the rate of some million acres a year. This means fewer orange and walnut groves in California, dairy farms in Georgia, truck and tobacco land in Connecticut, and less of the proverbially fertile valley land along the Miami in Ohio. All of these instances I have seen, as I have seen 15,000-acre tracts of the best farm land condemned for military installations when less productive sites could have been chosen.

There are some 500 major cities of over 25,000 population in the United States. Assuming that they could be evenly distributed, and neglecting smaller towns and cities, each would be in the center of a rectangle roughly 80 miles square. I have seen a fair number of them in recent years and recall very few that were not sprawling out into suburbs with little heed to open space, recreation, agriculture, beauty, or even the protection of future values. An exception, as a taxi driver profanely informed me, was not growing because the local university had everything sewed up!

Since this problem of urban sprawl is now receiving intelligent attention in a series of articles in *Fortune* magazine, I shall note only that it is serious, immediate, and far from simple. Municipalities generally have powers of expansion and taxation against which the rural landscape is without defense.

And between cities, across the land, highway departments are busily freezing the nation into a permanent interurban geometry. Often, in fact if not in theory, they are responsible to no one but themselves and their Euclidean rule that the shortest distance between two points is a straight line. Only through leaders who will devise and citizens who will support better use of urban and highway space can growing blight be checked. Professional planners, who, by the way, are seldom summoned until it is too late for them to be of real use, now frankly regard the entire strip from Washington to Boston as one great metropolitan area. Any lingering doubts on this score should fade at the sight of a new throughway blasting its course among rocks and homes, across land and water.

At Washington, southern end of the megalopolitan strip, fateful decisions regarding the future allocation of American space are made. One of the cabinet members who has much to do with such decisions told a recent visitor, "For one individual who, like yourself, comes here to protest the exploitation of wilderness areas, parks, and other public lands, there are a dozen who come here to press the opposing view." No matter what the sympathies of such a public official, these are elementary facts of political life with which he must reckon.

There are, moreover, numerous agencies of government, not always in close harmony, that are charged to administer space and its resources. What happens is the resultant of many forces, including the pressure put upon Congress and the advice it receives from appropriate bureaus.

The late Colonel Greeley used to relate how much of our national forest space was reserved. Congress, alarmed at the rate at which Theodore Roosevelt was setting aside forest reserves, lowered the boom on him, but the law could not become effective until he signed it. During the few weeks of grace Roosevelt, Pinchot, and Greeley spent evenings sprawled on a

White House floor with maps, for all the world like kids with a comic supplement, marking out forests while the President still had power to do so.

Unlike most cities, Washington was built on a definite plan and is still under close supervision. But the unremitting pressure of housing, traffic, and waste-disposal problems is a constant threat to the space required for recreation, let alone for esthetic values, traditionally a matter of concern. Among other things, the Potomac is notoriously polluted, and the pressure for schools and other public facilities in the overflow region outside the district is a headache to all concerned. In these environs, as around growing cities all over the country, one sees a wilderness of houses built to sell. And the buyer is usually more concerned with pushbuttons and gadgets than with sound construction. It takes no prophet to visualize what the condition of these potential slums will be in less than a generation.

Not quite midway to Boston is Jersey City and the whole complex of sleeping towns for New York. As of October 25th of this year the authorities of Jersey City were weighing the relative merits of pails versus paper milk bottles to dispense drinking water. The reserve for Jersey City and nearby places was then enough for about one month. Not even that flower of technology, the modern city, is exempt from the pressure of natural forces. Nor should this be surprising. While the per capita demand for water rises, so does the area that is waterproofed and designed to get rid of rain as fast as it falls.

Even the air is a problem. One approaches the Hudson through one of the most unsavory mixtures of gases on earth. What smells bad, with such noble exceptions as Limburger and Liederkranz, is seldom good. New York City, whose canyons full of fumes are no bed of roses, is within the same general zone of turbulence. The resulting uproar reminds one of the classical dispute as to which stank worse, a goat or a tramp.

New York City illustrates what might be called a space paradox. As its population has grown, so has the per capita space, except possibly in some very congested areas. At the same time, the rural areas, whose emigration supplied the growth of nearly all major cities, have fewer people. Farms are increasing in size; fewer men are farming larger farms. Everyone is getting more space while the population rises. The answer is, of course, that the rural man who becomes urban is not getting more space than he had—simply more than he would have had had he moved into the city a generation ago.

Further north, in New England, we encounter other interesting problems of space. Most of them involve conflicts of interest, often elements of minor tragedy. I have in mind the annihilation of homes for which money cannot compensate. One such, whose sturdy hand-made beauty, books, pictures, and furnishings represent the slow accretion of high cultural influence—not mere personal luxury—is now untenable because no better way has been found to dispose of the garbage of an expanding dormitory population than to burn it nearby. In southern Massachusetts I saw the occupants of a group of new homes trying to repair the damage of flood in a site which was notoriously subject to high water. The unwary newcomers who bought these houses did not know this and no one warned them.

It was, in fact, the floods of 1955 that revealed most dramatically what can happen when important fields of science are neglected while others are being applied to the limit in technological development. Manufacturers of electronic equipment, optical instruments, and precision tools certainly keep abreast of scientific developments. Yet in locating their plants they took risks which no geologist, or competent botanist, would have sanctioned had his opinion been sought. Not only did they expand their activities upon the hazardous flood plain, but in many instances they intruded upon the channel itself, thus making bad matters worse. The old water-mill builders took no such chances. Their homes were on high land, for they knew and respected the power of water.

New England, northern end of the great metropolitan strip, offers many other examples of the pressure of humanity upon space, although it has no monopoly in this respect. The West Coast, the most rapidly growing area in the nation, may be more graphic, for it lacks any protection from past cultural inertia. But in New England one may see a losing struggle to preserve esthetic and recreational values in the face of an insistent desire to expand industry, cater to the automobile, mine for gravel and rock, convert the rivers into free sewers, and in divers other ways capture the nimble dollar.

In these respects New England is no worse—and no better—than other parts of the nation. Two-thirds of its hinterland are now covered with forest, largely of poor quality, occupying land that was once farmed and later grazed during a booming wool industry. Yet this two-thirds of the area produces not more than ten percent of the rural income. In contrast to this, I know a Danish forest of 2000 acres that furnishes year-round employment to 50 men and 20 additional during the winter months, all at a profit. True, the New England soils are often thin and not highly fertile, but the chief trouble seems to be that we have consistently used up the finest trees, while the Danes since 1800 have been saving them for seed stock. Even though one cannot increase space, proper measures will greatly increase its yield. Inferior races of trees are just as wasteful of space as inferior breeds of livestock on pasture and farm. While New England forests even in their present poor condition add vastly to the beauty of the countryside, the time is not too far distant when their products will be needed. European experience shows that good yield is quite consistent with esthetic value.

Across the continent, with infinite variations due to local conditions, the problem of space is growing in urgency. Ultimately we shall have to face the purely physical fact of increasing numbers on a finite area containing finite resources. Of these resources, water is now getting some of the attention it deserves. But we should keep other substances in mind, recalling that we, with less than 7 percent of the world's population, are now absorbing more than 60 percent of the world's mineral production, or ten times our quota.

Meanwhile, the general pressure is complicated by conflicts of interest. Different groups and individuals see different possibilities in the same area, and all alike wish to secure the most from it. As great cities grow they become more, rather than less, dependent upon widening circles of rural land—for water, milk and other food, transport, recreation, housing, labor, and income. It is interesting to consider the sources of support for the four world territories that have more than 10,000 people to the square mile. They are, in order, Macao, a shipping and commercial center, Monaco, a gambling resort, the District of Columbia, where taxes are collected and spent, and Gibraltar, a military post! No great concentration of people is ever self-sustaining. The Valley of the Nile, which has had perhaps 1000 people to the square mile for millennia, depends upon the vast headwater areas reaching south to Lake Victoria for its water and fertility. The same principle applies to the crowded downstream river margins of China and India.

The time must come when we shall have to deal openly, honestly, and realistically with the basic biological fact that numbers of organisms cannot multiply indefinitely within a finite area. And since our own species is under discussion, we must face the unparalleled conditions of increasing numbers and biological dominance combined with accelerating mobility, power, speed, and consumption on the part of the individual. Eventually we must come to grips with these fundamentals. Meanwhile we can, in my judgment, help matters greatly by admitting that conflicts of interest do exist, identifying them, and establishing some order of priority for conflicting claims.

I have no easy solutions to suggest. The first step in dealing with a scientific problem is to make it clear. This I have tried to do, aware of the fact that in our society solutions must be worked out by common consent—generally a painful process. There is a maxim among medical men that more mistakes come from not looking than from not knowing. So far as space is concerned, both looking and knowing are involved.

Much concern is now being expressed for better science training. Here at least we can make a sound, if modest, beginning. Training in the rudiments of science—asking, observing, and reasoning—should begin along with training in the mother tongue and be a part of the same process. College science, training as it does both scientist and citizen, should be taught in context with the rest of human knowledge and experience. It should certainly be a convincing and challenging aspect of education. How far it falls short of these ideals one can discover by asking those who have been exposed to it. Always excepting those who have an innate taste for science, the average college graduate, in my experience, does not retain enough for literate conversation upon the subject, let alone enough understanding to use it in civic affairs. Too often his mood is one of active distaste.

As a rule he has been required to take *a* course in *a* science weighted too often for the benefit of those who must go on in the particular field. How many times I have been told by colleagues: "We must teach it this way, or our students will not be ready for the next course." Such a philosophy misses the fact that by sacrificing insight to detail, fine intellects that might be potential candidates for further work may be lost.

Nor does the mischief stop there. No one science by itself can give that balanced view of the world of nature so essential to the citizen in our modern culture. A peep-show, no matter how good, is no substitute for a panorama. Until citizens, administrators, engineers, and businessmen become aware of the broad sweep of science, we may expect to see it applied, as it has been so largely, for immediate return rather than ultimate and lasting benefit.

Let us, therefore, avoid the folly of thinking that science can be separated from the broader problem of education and self-discipline. The present hue and cry for more and better science education could easily lead us into the trap that caught the Germans in 1914. More and better science teaching we must have, not merely to produce needed scientists, but to create an atmosphere of scientific literacy among citizens at large. Only by general understanding and consent can truly creative science be sustained within our system of society and its results applied for the ultimate welfare of mankind.

Liberal education today should require not less than two years of college science, based on a skilfully planned and interwoven sequence dealing with time, space, motion, matter, and the earth and its inhabitants. Nothing less than this is adequate for a proper appraisal of the natural world and our role as a part of it. This experience should be obtained at the hands of men who believe in it, who have status with their colleagues, and who are in intellectual communication with each other. There is no place for loose ends or superficial business in such an enterprise. Nor can it be carried on without the actual contact with phenomena in laboratory and field. Science that is merely verbalized is dead stuff.

But to this end it is equally essential that the educated individual must acquire such experience in the context of history, the arts, and an understanding of his own species. As a rough objective, I would propose turning out a product aware of what is going on around him in the world of nature and of man, able and willing to relate the present to the past and to the future in both thought and action.

To do this we must recognize with greater frankness than we have that there are vast differences among individuals. Let us learn to look upon these differences with respect, as a source of enrichment rather than discrimination, training each, honoring each, and expecting service from each according to his gifts. Let not the slow impede the fast, nor the fast bewilder and condemn the slow.

With a population set to double in less than half a century, with a national space which, though vast, is finite both in area and quality, with each individual making growing demands, moving faster and further by a factor of at least ten, we have on our hands a problem without precedent in geological history. But if we sense the problem and believe it worth solving, we can solve it.

Our future security may depend less upon priority in exploring outer space than upon our wisdom in managing the space in which we live.

The privileges enjoyed by the scientific-technical community before Sputnik were dwarfed by the attention paid it afterwards. Economy-minded, the Eisenhower administration had intended to reduce certain science funds. Secretary of Defense Charles Wilson's view that "Basic research is when you don't know what you're doing" was intended to be reflected in lowered support for it in the future. The Air Force Office of Scientific Research had been notified that its budget was to be severely reduced; after Sputnik, however, its budget was expanded by almost 40 percent. New institutions were created to deal with the Soviet space menace. The President's Science Advisory Committee was created, and James Killian of MIT was named the first Presidential Science Advisor. A space agency [the National Aeronautics and Space Administration (NASA)] was quickly created out of aeronautic predecessors and immediately became a major sponsor of research. Expenditures on medical research, too, were dramatically increased during Eisenhower's second term. Not only was the Space Act of 1958 passed, but the National Defense Education Act provided new support for science and science-related education.

For fiscal 1959, basic research was supported in relatively munificent fashion by NASA (27 percent), the Defense Department (22 percent), the Atomic Energy Commission (16 percent), and the National Science Foundation (10 percent), among many other agencies. By fiscal 1960, the total federal budget obligations for research and development reached over $8 billion, more than $6 billion of which came from Defense and $425 million from NASA. By 1963, obligations soared over $13.6 billion (Defense providing $7.3 billion and NASA over $3.4 billion). Expenditures on research and development reached 12.4 percent of the overall federal budget.

Although outlays for basic research continued to be much smaller than those for either development or applied research, money—even for pure science—seemed to be plentiful. The prestige (and fear) commanded by the scientific enterprise went largely unchallenged. The social sciences, as well as the physical and biological sciences, reaped the benefits. By 1967, the Department of Defense inaugurated Project Themis to foster new centers of academic excellence. At the request of the Defense Department, the National Academy of Sciences convened a study group to advise it on areas of social science that could reasonably promise "great payoffs" within ten years. The panel recommended increased Defense support for and use of social science research.

The increased federal funds brought increased burdens. In his presidential address of 1959, Wallace Brode underlined the necessity of a systematic approach to science policy. "[A] crisis has been reached," he said, "in [the] overlapping expansion and growth of our science programs, so that none of these major programs can be adequately supported except at the expense of the less glamorous areas of science, education, and culture, which are, nevertheless, essential to our basic welfare." With the federal government supporting about 85 percent of basic research, and with defense-related agencies providing the lion's share of that, serious questions were raised concerning science's independence and its future support.

While many influential statesmen of science relished the anarchy (or perhaps in their view, variety) of sources of support, Brode insisted that the scientific community examine "a more orderly way of funding," and warned of Eliza Doolittle's fate in *Pygmalion*—that is, being cast adrift after being raised to a high level of competence. Brode suggested the possibility of a cabinet-level department of science to oversee a national science policy and the possibility of evolving a systematic means of employing funds for university research.

Paul Gross documented, in his 1963 address, what he termed "the greatly expanded tempo, scope, and power evident in the development of science." He noted once again the importance of science and technology as instruments of national policy and concerned himself with implications for science of highly organized, amply financed research. By 1963, money for science and technology was a considerable and noticeable drain on federal resources. Gross stressed that, given the "almost exponential growth of science and technology . . . it is obvious that some adjustment in the growth rate of federal expenditures for science must take place." Like Brode, Gross wanted a system of project priorities. Like Sears, he was worried about the ability of science and technology to alter the environment. Gross looked to the AAAS's traditional purposes: it was the duty of the scientific community to aid in the public understanding of science and to promote human welfare.

Alan Waterman's address the following year continued the discussion. While acting as a spokesman for more funds for "free" basic research— Waterman was director of the National Science Foundation—he also cautioned scientists about deleterious effects: "too much assistance to thesis-writing graduate students, with an eye toward grant or contract renewal; hasty writing and issuance of research reports scanty on detail and acknowledgement; a tendency to keep a weather eye on funds for extra salary or other perquisites." But, after all, Waterman remained, as did most of his colleagues, optimistic: "[W]e stand at the threshold of scientific findings that will pave the way for developments of a different order of magnitude and novelty than the world has ever known."

Development of a Science Policy

Wallace R. Brode
Chicago, 1959

Wallace Reed Brode (1900–   ), a chemist, served in the Army in 1918 and, after graduation from Whitman College (Washington) in 1921, continued his studies in organic chemistry at the University of Illinois, Urbana (Ph.D., 1925). He worked as a research chemist for the National Bureau of Standards and, after two years of study in Europe as a Guggenheim fellow, joined the chemistry department at The Ohio State University in 1928. His special fields of interest—spectroscopy, dyes, and applied optics— were reflected in the laboratory program he developed at Ohio State and in his pioneering textbook on chemical spectroscopy. During World War II, Brode acted as a consultant and worked on projects for the Office of Scientific Research and Development in London and Paris. He also worked on the ALSOS Mission for the capture and interrogation of enemy scientists.

Brode resigned from his academic position in 1948 to become associate director of the National Bureau of Standards. For the next ten years he supervised chemical research and was responsible for the bureau's education, publication, and foreign relations activities. Appointed special scientific adviser to the Secretary of State in 1958, Brode directed the revived U.S. scientific attaché program abroad. He served as president of AAAS in 1958.

"The liberal spirit which animates both Congress and the executive departments in their dealings with scientific affairs is very apt to lead them into the support of scientific enterprises without any sufficient consideration of the conditions of success and of efficient and economical administration; and a careful consideration of each proposed undertaking by a committee of experts is what is wanted to insure the adoption of the best methods."

These words appeared in an editorial in *Science* published 25 April 1884. The magazine *Science* was then only three years old, the American Association for the Advancement of Science was 36 years old, and there were only four or five government agencies active in science. What prompted this editorial of 75 years ago were the lively discussions on the need for a Department of Science in the government. Then, as now, the scientists as well as other government officials were divided in opinion. In 1884 Congress appointed a commission, known as the Allison Commission, to consider the creation of a Department of Science. The President of the National Academy of Sciences, O. C. Marsh (a former AAAS president), was asked to name members of the Academy to serve on a committee to assist the Allison Commission. Among the members named were two distinguished scientists, Simon Newcomb and Cyrus Comstock. This committee was to survey and study the procedure of handling science in other countries and to recommend methods of coordinating the scientific areas. Simon Newcomb had served as president of the American Association for the Advancement of Science seven years earlier, in 1877, and was one of the Navy's most illustrious scientists. General Cyrus Comstock was an equally distinguished academician who served in the Army. However, the Secretaries of the Army and Navy both objected to a government scientist from their agencies serving on a National Academy of Sciences committee which was to advise the government.

One of the most outspoken in favor of a Department of Science was Major John Wesley Powell, a vigorous and colorful government scientist, who was chief of the Geological Survey, a non-military establishment. Powell appeared on 16 occasions before the congressional committee. He commented on the elimination of Newcomb and Comstock from the Academy committee and noted that the "military officer plans and commands; the civil officer hears, weighs and decides," and that "the military secretaries did not desire to have their subordinates deliberate on questions of policy affecting the conduct of the secretaries themselves." Powell, however, felt this suppression was justified in the military circles but would not have been justified in a civilian area.

The National Academy of Sciences committee expressed the feeling "that the time is near when the country will demand the institution of a branch of the executive government devoted especially to the direction and control of all of the purely scientific work of the government." However, if establishment of a department could not be effected, they felt that a coordinating scientific commission would be in order. Neither a Department of Science nor a Science Commission was established, due to various factors, including political changes in a new Congress.

Powell supported a Department of Science, although he favored some modification of the proposals of the Allison Commission. Powell was elected president of the American Association for the Advancement of Science in 1888, and I had thought that his presidential address before the association in 1889 might add some comment, in retrospect, to his testimony of four years earlier on a Department of Science. However, Powell's presidential address, published in the American Association for the Advancement of Science's proceedings, was a scholarly ethnological discussion on "Evolution of music from dance to symphony"—a subject on which he, as founder of the Bureau of American Ethnology and a student of American Indian culture, was most competent to address this association.

The very fact that the organizational issues being considered 75 years ago are basically identical to those of today raises this fundamental question: Has science changed in relative importance over the period of years since the founding of our country? Jefferson, as Secretary of State in our first cabinet, was in a sense also Secretary of Science because he handled such areas as patents, decimal coinage, and our standards of length and weight. Each major scientific or technological development has in its period of history created startling and revolutionary changes in the daily pattern of life. The telegraph as compared to the pony dispatch probably represented a more radical advance than did the telephone over the tele-graph, the radio over the telephone, or television over the radio. The electric light had just been introduced in 1884, and there appeared in the literature pronouncements about the "impact of science" which are as current as if made today.

Although it would appear that problems which existed earlier are still with us today, they are actually compounded. Whereas Thomas Henry, Asa Gray, Wolcott Gibbs, and others in the past produced their discoveries and theories as individuals, many of today's advances represent the work of teams, whole laboratories, and industries. This complexity in the scope and size of scientific operation will not remain static but will continue to increase dynamically.

Hence the concern of the scientific community is justified. If one feels that present arrangements are unwieldy enough today, what will their state be in another 100 years?

The American Association for the Advancement of Science has led in exploring these problems through its Parliament of Science, its Basic Research Symposium, its regular meetings, its programs, and its publication *Science*. These have all helped to stimulate the consideration of our science policy. In the past two decades there have been a number of governmen-tally appointed boards, panels, and commissions to evaluate the place of science in our nation and government. Nearly 15 years ago Vannevar Bush, as the coordinator of the nation's defense research, issued his famous treatise *Science, the Endless Frontier*. In 1946 the President of the United States created a President's Scientific Research Board, under the chairman-ship of John R. Steelman, to study science and public policy; and in 1947 this board issued a four-volume work, known as the Steelman Report, under the title *Science and Public Policy*. Six years ago the Commission on Government Reorganization, also known as the Hoover Commission, was concerned with our expanding government. In 1956 the President's Com-mittee on Scientists and Engineers was created; in 1957 the President's Science Advisory Committee was established, and upon the recommenda-tion of this body in 1959 the Federal Council for Science and Technology came into existence.

All of these groups have devoted considerable time and energy to various phases of government operation and the role of science. The National Science Foundation is a direct result of recommendations made by Bush and the Steelman Report. As a result of the Hoover Commission studies there evolved the Department of Health, Education, and Welfare, which encompasses major social and health agencies, but whose ultimate formation was not accomplished without considerable controversy and discussion.

A recent editorial in one of our nation's leading papers notes that chairman McCone of the Atomic Energy Commission "is convinced that the United States—if it is not to become technologically and economically inferior to the U.S.S.R.—must work out methods of marshalling its scientific and technical talents for concentrated *top priority* work on projects of overriding significance." McCone further proposed that this increased activity in a specific scientific field should be made "at the expense of projects of lesser importance." If one did not know the scientific subject matter of chairman McCone's agency, one might well ask, "Is this space, oceanography, undersea geology, meteorology, medicine, education, high-energy physics, atomic energy, food, materials, weather and smog control, or transportation?" Each of these areas has supporters who feel that programs varying from ten million to ten billion dollars a year are essential to scientific progress.

This *overriding* or *top-priority* attitude of some specialists is a reflection of the enthusiasm for one's own field of specialization. However, a crisis has been reached in this overlapping expansion and growth of our science programs, so that none of these major programs can be adequately supported except at the expense of the less glamorous areas of science, education, and culture, which are, nevertheless, essential to our basic welfare.

Why have we arrived at this state of crisis? The answer is *expediency*. The immediate conditions and circumstances existing at the time determine which programs are "top priority." Even when the stimulating conditions have been removed, the top-priority label is often maintained. Then another set of circumstances dictates the creation of another top-priority label assignment to another area of science.

Even though "atoms for peace" is slowly being transformed into "science for peace," there do exist separate areas of nuclear and atomic research which enjoy an elevation high above science in general. Many less well developed nations have devoted time and resources to nuclear and atomic programs rather than basic science. There are areas of the world where basic education, health, and agricultural training should predominate, but we find these countries building nuclear reactors by crude methods of hand labor. When we ask what is to be done with the reactors, we are advised that they are training reactors, to train people to run more reactors.

The latest area of science which is capturing the minds and purse of our nation is space. There are a number of good contenders in the race for future top-priority assignments. Perhaps the next will be oceanography, weather, or materials research. We do favor advances in these areas of science; we need this progress, but this progress should not be effected through a corresponding reduction in the rate of advance for other, equally important areas of science.

How can projects which are "overriding" and those which are of "lesser importance" be identified? Who is to make this allocation of relative effort? One of the most difficult tasks facing us is to achieve a long-range planning effort which would remove expediency as the sole controlling factor. A national science policy is needed for a wise and rational distribution of scientific activities, so that space, defense, education, atomic energy, oceanography, and medical research are not bidding against each other for limited available support. The growing demand for scientists in the face of a limited supply of scientists, materials, funds, and facilities requires major policy decisions as to the distribution of resources. These decisions should of course include the extent to which specialized agencies may recruit by scholarship, fellowship, and research support.

Every enthusiastic scientist with a dream for the future can envisage space ships at his command; areas of flashing lights and computing machines reading, translating, abstracting, and digesting the world's literature, even solving the problems punched into the machine; or reflecting radio telescopes a mile in diameter to enable him to communicate with other worlds. However, there must be a limit, and not only must scientists realize that there should be a relative priority assigned to areas of science but there should also be recognition that scientific programs do not all have priorities that override economic, political, educational, and social developments.

With tongue in cheek, a past president of the American Association for the Advancement of Science, Warren Weaver, approached this problem of super-programs and priority assignments in a recent article in *Science* [**130**, 1390 (1959)], in a clever satire on the "Report of the Special Committee." The "Weaver Report," along with Parkinson's Law, may provide the means for arriving at some solution to our problem, by taking a distorted view and working backwards toward a rational solution.

We have seen a marked shift in the responsibility for the support of scientific research in the past 30 years. In the 1930's the government was supporting only about 15 percent of the nation's basic research, which was almost exclusively in its own laboratories. Today, the federal government is supporting about 85 percent of the nation's basic research, of which still only about 15 percent is in its own laboratories. This great growth, in both percentage and total amount, has been primarily in research support in industrial and educational contracts. Today our government's total budget is many times larger than the budget of the 1930's, and four-fifths of this budget is for defense activities and less than one-tenth for normal governmental activities. The expenditure of 1 percent of the defense agencies' budget (and I am including the area of applications of atomic energy and space research as defense) in support of research means about $500 million a year, which is about 85 percent of the nation's program in research. This means essentially that our scientific research program is directed and guided by these agencies.

It is certainly true that after World War II there was an emergency situation and that without the aid given our educational and research programs by government agencies concerned with defense and applied sciences we would be in a sorry mess, unless—and this is the great unknown—the pressure of the situation would have been sufficient to create a more logical civilian support to this civilian activity. It is generally agreed that the initial "bailing out" of universities by the opulent agencies required only a small portion of their total budgets and was a good thing in the immediate situation. Just how much this expedient delayed general support to science education, coordinated science programming, or a national science policy is unknown and may never be clearly recognized.

The Steelman Report recommended that the National Science Foundation should support basic research in governmental establishments as well as in universities, yet the pattern of today's grants is essentially to universities and does not combine what could be the government's broad interest and coordination of the entire national science program.

In many of the basic-science areas of the government, such as the National Bureau of Standards, the Weather Bureau, the Geological Survey, the Bureau of Mines, and the Forest Products Laboratory, which are attached as appendages to major departments, personnel supported by direct appropriation are essentially at the prewar level of the 1930's. The growth of the bureaus needed to keep up with our expanded science program is almost entirely dependent, as have been our universities for support in research, on contracts from the large agencies concerned with applied sciences and defense. Another justification for the continuance of research support both in universities and in nondefense government laboratories by the major agencies concerned with defense and applied sciences has been that, with multiple supporting agencies, contractors can shop for different sponsors. Since this research is only a minor function of the supporting agency and often is not directly related to its mission, less coordination might be expected than if all funds came from a single agency whose principal function was to support our national science and educational program.

In the establishment of the National Science Foundation many of us who followed the discussion of its formation felt that the nation's responsibilities in basic research would, to some extent, be absorbed from other government agencies, concerned with applications, which had assumed some of these responsibilities. To prevent a drastic movement of the support of basic research to the National Science Foundation, it was actually indicated that the various agencies concerned with applications should continue to support, in their own organizations, "basic research in areas which are closely related to their mission."

There was no intention of removing basic research from agencies and laboratories which require it in their developmental activity. In fact, the thinking scientist has often been worried to find that there did not appear to be a sufficient guarantee of adequate interest and support of necessary basic research in the agency establishments concerned with applications.

Most of our technological agencies have special authorization in their enabling acts permitting them to engage in basic research in support of the objectives of their mission. So far as I know, there has been no question raised as to whether such research was proper. However, many technological agencies have used such a research authorization to justify their support of research programs in all areas of science rather than in those areas which would appear to be directly concerned with their proper mission. Even the National Science Foundation, which has a rather broad mission concept, is aware that certain phases of science, such as medicine and agriculture, are in the areas of responsibility of other government agencies.

Nearly all of the agencies concerned with applications have indicated that they feel a responsibility to help train the research workers of tomorrow through scholarship and research grants. If such agencies strongly felt a responsibility, they would show no reticence in providing such funds to the universities as university grants, or in transferring these funds to the National Science Foundation to augment a planned national program of general support. What is disturbing is an insistence by each agency that its granting office should be able to select and direct both the recipient and the subject of research so as to exercise a guiding hand in our educational and research institutions. These amounts are not small, for they constitute much more than half of the total support to basic research in this country.

The problem which concerns us is that Army, Navy, Air Force, and space and atomic energy agencies have had to assume responsibility for contracting for this very extensive amount of basic research in universities, foundations, and government laboratories, both here and abroad, in all phases of science. While most of these supported groups and many scientists in this country do not question the source of the "collection plate," it has become more apparent to me in dealing with foreign science programs that there is hesitancy abroad on the part of scientists or universities at becoming involved in programs supported by a foreign military agency. This is especially so where the country is essentially neutral or where the program is not a part of a mutual defense act in which the scientist's own country is participating and to which the proposed program is attached.

With the great increase in recent years in the support of basic research by public agencies and foundations and the predicted doubling of research in the next ten years, there is reason to feel that a more orderly way of funding might be evolved than reliance on a multiplicity of sponsors. The urgency of the need for a revision of our program methods becomes greater as the size of programs increases so as to involve large sums of money, large numbers of individuals, and many research facilities. The pressures for priority action on massive programs, such as those concerned with space ships, new sources of energy, or weather control, often develop in the nonscientific political or economic areas and work their way towards the university, industrial, and government research worker.

Many of us are concerned, as was Eliza Doolittle in Shaw's *Pygmalion* (and in *My Fair Lady*), who plaintively asked "What's to become of me?" when, after being raised to a level of competence and ability, she was threatened with the possibility of being cut off from her subsidized support and faced the prospect of being heaved out on the streets and having to shift for herself again.

The government has a responsibility, which will certainly grow in size and scope, to support a major share of the nation's research and applied-science programs. This responsibility in the area of advanced scientific education and basic research in universities, institutes, and government laboratories must be fully met if we are to maintain a technological leadership in the world.

There should be a revision and realignment of our support so as to provide more direct and less controlling support to universities, and greater direct support to government basic-science programs in the government's own laboratories. Serious consideration should be given to the reduction or elimination of "convenience" or synthetic scientific agencies which, while operating as non-government laboratories or institutions, are doing almost exclusively governmental work with government funds. These laboratories should be made bona fide government laboratories, and they should be so directed and set up as to permit the government to do its own essential scientific work under working conditions which are most conducive to efficient and effective operation.

I feel that the government does have a responsibility to continue to support science, just as it has a responsibility for health, agriculture, and defense. There must, however, be some instrumentality with a considerable degree of control, which can decide when to support, when to taper off, or when to terminate various research programs—and such responsibility must eventually center in a coordinating establishment such as a Department of Science.

If we are to maintain in the government high-level policy and research positions for scientists, we must provide compatible employment and a challenge to their capabilities in the responsibilities assigned them. It is difficult to attract or hold good scientists if there is no future level of administrative or research responsibility which they can expect to reach. Much of the government funds for applied research for the government are used to maintain non-civil-service laboratories or organizations such as Los Alamos, ARPIA-IDA, Lincoln, Brookhaven, Oak Ridge, the Applied Physics Laboratory, and many industrial laboratories. These quasigovernmental laboratories provide higher salaries and better working conditions than are provided scientists in the government and—perhaps most important to the scientist—provide high-level, polity-type positions of responsibility, and the employees are treated as non-governmental scientists in dealing with the government. It would seem that a number of these contract operations in science and technology should be reexamined to determine whether there should be an improvement in the status of the government-employed scientist and direct absorption of much of this work in a Department of Science or other coordinating science structure in our government.

One interesting recommendation of the Steelman Report was that the board of the National Science Foundation should be composed of "distinguished scientists and educators to be drawn one-half from the Government and one-half from the outside." Actually the board includes no government scientists but is made up exclusively of university and industrial leaders, with university personnel predominating. In selecting personnel to run government departments, bureaus, or offices it is generally the procedure to try to find someone to bring in from the outside. Usually such a person is on leave of absence from his industry or university for one or two years. Seldom is the appointment of a career civil servant seriously considered, even though there are in the government career civil servants who may be recognized in other nations as world authorities and leaders in the field.

There should be established a policy whereby top management is not maintained on a rotational system by persons who are not directly associated with government operations. It is certainly true that many a highly respected and competent university professor, lawyer, or businessman has at considerable sacrifice of time and money, and often of prestige, agreed to come to Washington for a one- or two-year period to pitch in and help run the government. Nevertheless, the government should develop within its own establishment sufficient capability to operate its agencies, and our policy decisions, whether they be science, taxation, or welfare, should not be made without guidance from personnel experienced in governmental operations.

The problems to be faced are these: (i) determination of the direction in which science will advance and of the areas in which continuing or new programs are to be supported; (ii) the emphasis and relative priorities to be placed on scientific programs, including not only the "top-priority" programs but also the minor programs which need to be kept alive and operating on a modest scale; (iii) the administration, financing, evaluation and support of our science programs within the government; and (iv) the distribution of responsibility for the carrying out of scientific programs between government laboratories and university, private, industrial, domestic and foreign (intergovernment, government, and private) facilities.

The direction in which science should be encouraged to advance will in part be spontaneously determined for us by the inquisitive research worker who has been given freedom to probe. Essentially this is exploratory research and should be supported by institutions, so that research workers may spend a reasonable portion of their time in exploration. In addition to this limited area of free research there is our major area of programed research, for which there should be some over-all plan. Such a plan must establish the relative priorities or emphasis and the rates at which certain programs should be pursued.

Our problem today is that we have reached a saturation point with respect to available personnel. Hence, further expansion or support in many fields must of necessity require reduction of the active available material in other science programs. Such a disturbance in our present unstable equilibrium of distribution of effort is particularly felt when massive new programs are initiated, such as new billion-dollar efforts in space, oceanography, or health.

The Hoover Commission noted that our government is expanding to such an extent that the executive branch has 74 agencies "which divide responsibility and which are too great in number for effective direction from the top." The commission recommended that certain of these should be "grouped by related function under the heads of departments." They indicated "in many cases several agencies each have a small share in carrying out a single major policy, which ought to be the responsibility of one department." In summary they concluded that the government must "create a more orderly grouping of the functions of government into major departments and agencies under the President."

The Steelman Report in discussing the creation and operation of a proposed National Science Foundation suggested that the Foundation "should be located within the Executive Office of the President until such time as other federal programs in support of higher education are established. At such time, consideration should be given to grouping all such activities, including the National Science Foundation, in a single agency." The National Science Foundation, which was subsequently established, has a more scientific than educational mission, and with the establishment of other new science agencies, such as the space agency, it would appear that the Steelman recommendation might well be applied in principle toward the creation of a Department of Science.

There is little if any opposition to the broad concept that as scientific or any other activity grows in size and complexity and begins to be a major consumer of personnel funds and facilities in our economy and culture, some consideration should be given to increasing efficiency in the utilization of our limited resources through suitable coordination and planning.

The President's Science Advisory Committee, a group of nongovernmental scientists, in viewing from outside the government the problem of science in government, suggested that there should be created some instrumentality to promote closer cooperation among federal agencies in planning and managing their program in science and technology. They recommended the establishment of a Federal Council for Science and Technology, and the President issued an executive order accordingly, in March 1959. This council is an inside-the-government group consisting of policy members of departments or agencies concerned with science. The representatives, however, need not be scientists, and as the council is presently constituted, many are not scientists.

It would seem reasonable that the Congress should be able to seek advice and counsel from a coordinated science leadership in the government. Early this year the chairman of the congressional committee held hearings on a possible Department of Science and sought the appearance as a witness, for advice and comment, of the chairman of the newly created Federal Council for Science and Technology. The council chairman declined to appear, on the grounds that he was a privileged member of the President's staff, yet as head of the Federal Council for Science and Technology he was responsible for the organization which was charged with making recommendations for the creation of effective means of promoting a more efficient, coordinated science program in the government. It was my own opinion, which I expressed when, as president of the American Association for the Advancement of Science, I was asked to appear before the same congressional committee, that a strong and responsive Federal Council for Science and Technology might well evolve into a Department of Science.

The inability of the Congress to draw on the advice of the existing Federal Council for Science and Technology is in itself an indication of the need to separate our science coordination direction from the President's own office and Science Adviser so that the coordinated leadership in science in the government may speak with the authority of the group it represents rather than only through the President. I feel that the President should continue to have a strong science adviser, however, in the presentation of a science program for the government. The leader of the program should be able, like the heads of other agencies, committees, departments, or commissions, to speak for his agencies before the Congress, and also to report the findings and recommendations of those agencies to the President. It would seem that in the organization of science and our governmental science policy, as in the areas of labor, commerce, the military, health, education, and agriculture, it should be possible for the planning and organizing committees of Congress to have reasonable access to the agency in question for advice and assistance.

The President's Science Advisory Committee considered the suggestion of a Department of Science as a proposed solution to the problem of coordinating the nation's science programs but defined a Department of Science as a department which would bring together "*all* of the government's numerous scientific and technological operations," including the scientific phases of defense, agriculture, and health which were directly related to the missions of the agencies with responsibility in these areas, and pointed out that such operations "could not be satisfactorily administered by a department far removed from the problems that are to be solved." This is a synthetic defense, in that elimination of research or development work pertinent and essential to the proper missions of these agencies was not proposed or recommended by the promoters of the legislation actually before Congress. The proposed legislation for a Department of Science was somewhat nebulous, but its main purpose was to stimulate discussion, and its originators recognized that it did not present a Department of Science concept that would be fully acceptable to all.

I consider as not of significance the arguments presented in the congressional hearings in which those who opposed a coordinated science program maintained that such a program would place too much centralization of authority in a single agency and establish too much government control. Neither argument is tenable if proper administration is provided and safeguarded.

A corrupt and mismanaged program would be bad, but to argue against any coordination on the grounds that it *could* be bad is just not logical. We should favor a governmentally supported program and at the same time incorporate into such a coordinated program the essential safeguards used in our local, state, national, and international institutions, both governmental and private.

Belgium, France, South Africa, and England have created cabinet or semicabinet posts for science departments. England has already indicated the nature of agencies to be absorbed into such a collation. These include space, atomic energy, health, research grants, and specialized science agencies such as standards, weather, patents, and science information. In his first press conference Lord Hailsham, Lord Privy Seal and the new Minister for Science in the British Cabinet, pointed out that "whether or not there is a need for a Minister or Ministry [in science] . . . there is a need for a policy in science and that policy cannot be a product of government thinking alone." Lord Hailsham emphasized that his Advisory Council on Science Policy "provides one of the keys to the present situation composed as it is of a unique connexion of Government and non-Government scientists. . . ."

Pioneering in the creation of a Department of Science in a government took place in this country about 100 years ago, under somewhat unusual circumstances. In the process of reviewing material for this presentation I became interested in how many presidents of the American Association for the Advancement of Science had been government scientists for a considerable portion of their careers. I found that about 30 out of the 112 had been government scientists. One of the former presidents of the American Association for the Advancement of Science was John L. LeConte, in 1874. In looking up his background I was intrigued to find that he had two cousins, Joseph LeConte and John LeConte, who had served as members of the Science Department of the Jefferson Davis government during the Civil War. Joseph LeConte also became president of the American Association for the Advancement of Science in 1892.

If we accept the concept that we need a national science policy to guide our scientific effort, and if, as a result of suitable commission and congressional action, formation of a Department of Science is proposed, we should be certain that it is in fact as well as in name an operating department. It should not be a superstructure imposed on existing organizations, but it should represent an honest and real effort to mesh the scientific interests and objectives of our government in the fullest possible utilization of resources. Thus, a Department of Science, while not removing from agencies such as Defense and Agriculture, concerned with applications, the research programs specific to their missions, should include all major segments of science not specifically pertinent to those missions. It should have separate bureaus or institutes with suitable directors of distinction to deal with space, atomic energy, medicine, weather, patents, science information, physical science, geology, and other recognized areas of importance. Each director should be aided by an advisory panel of experts in his area, drawn from academic, industrial and government sources.

To provide the Department of Science administrative head with broad and helpful advice it would seem reasonable to create an advisory council, which might be designated a National Science Council. Such an advisory group should not be exclusively academic but should include representation from government, science, and industry as well. Its principal responsibility would be to provide the Science Department administrative head with broad advice which might be helpful in arriving at decisions on the extent and character of support which the government should provide both to science programs in the government and, through contracts or grants-in-aid, to industry and universities—in short, the implementation of a National Science Policy.

These comments of mine on the creation of a Department of Science and a National Science Council have been postulations based on the needs that exist to coordinate governmental science, to create a National Science Policy, and to establish a liaison between governmental, academic, and industrial science. Before firm and thorough recommendations can be made, a commission should be established to study all of these questions very seriously. Such a commission should include representatives of government and of the academic and industrial community in both scientific and non-scientific areas. If the United States is to achieve a balance not just in its budget but also in its scientific programs, both immediate and long-range, there has to be a thoughtful and penetrating analysis of the problem. There must be the maturity of judgment and the courage of action required to change existing institutional procedures or philosophies where it is necessary. Where necessary there must be a facing of the problem and recommendation of essential drastic action, so as to avoid continuance of methods or actions which merely postpone the day when action must be taken.

The retiring presidential address before the American Association for the Advancement of Science is a personal presentation of my own ideas and is not intended to present the opinion of the association. . . . When a university professor presents a paper before this association it is taken for granted that his opinions are his own and not those of his university. In my presentation I have pressed for a more academic status for the government scientist, with opportunity to present for discussion and consideration before his fellow scientists his own concepts pertinent to science. I realize that there are many, including some members of the association's board of directors, who do not fully support or entirely share some of the ideas contained in my presentation.

In conclusion I would argue that some form of commission or study group should be established to give careful consideration to the problem of organization of science and science policy in the academic, industrial, and governmental areas of the nation, and that in this study serious consideration should be given to the following concepts.

1) There should be a regrouping of some of the government's scientific agencies or activities: either a Department of Science, a National Science Institute, or some other coordinated structure. A well-developed coordination must be established between the regrouped combination and those scientific agencies which remain separate, so as to insure an efficient and comprehensive National Science Program.

2) There should be a realignment of the distribution methods and responsibility for support of basic research in our educational institutions, with a movement toward university grants, administered largely by a department concerned with basic research, rather than by agencies concerned with applications. This may well need to be coordinated with the growing problem of support for our advanced-education program in all areas.

3) There should be some separation of governmentally sponsored, major research institutions from our educational and industrial system, especially of those institutions which are essentially concerned with applied science. There should be a greater acceptance of the idea of operation of such institutions under an improved, directly governmental administration.

4) The liaison of scientists in government with scientists in the academic field and in industry should be represented by a National Science Council in such a manner as to be compatible with the maintenance of our broad culture and balanced development.

The Fifth Estate in the Seventh Decade

Paul M. Gross
Cleveland, 1963

Paul Magnus Gross (1895–), *a physical chemist and educational administrator, helped effect the transformation of Trinity College into Duke University. Born and educated in New York City, Gross studied chemistry at the College of the City of New York and received the B.S. degree there in 1916. His graduate work at Columbia University was interrupted by service with the U.S. Army Chemical Warfare Division in 1918. Gross returned to Columbia and was awarded the Ph.D. in 1919, in part for his thesis on aspects of solution chemistry, a specialty in which he continued to do research throughout his professional career.*

At Duke, Gross conducted research concerning the fluorination of organic molecules. In the mid-1920's Gross investigated the biochemistry of tobacco plants—of critical importance to the economy of North Carolina. Gross spent 1929 at the University of Leipzig studying molecular structure with Peter Debye; Gross's investigations later resulted in a critical test of the Debye-Hückel solution theory. During World War II, Gross developed a frangible bullet for gunnery training; for this service, he received the government's highest civilian award, the Medal of Merit.

Postwar activities for Gross have included government consultant work and university administration. As the small college expanded into Duke University, Gross emphasized the quality of faculty, a policy he was able to implement in his positions as dean of the graduate school (1947 to 1952), dean of the university (1952 to 1958), and vice-president of the education division (1949 to 1960). Gross has been president of the Oak Ridge Institute for Nuclear Studies since 1949, is a fellow of the American Physical Society, and was president of the AAAS in 1962.

. . . Tonight, instead of talking of my own field of physical chemistry, I think it may be of interest if I say something of the present status of science and of scientists, as I see it, from my experience of almost a half century in scientific education, research, and administration. The reference in the title to the "seventh decade" is, of course, obvious. That to the "fifth estate" may not be familiar to some. This relates to the three estates of English history—the Lords Spiritual, the Lords Temporal, and the Commons. To these was added a fourth—by Edmund Burke, according to Carlyle. Burke is said to have observed, in a famous speech: "There were Three Estates in Parliament; but, in the Reporters' Gallery yonder, there sat a *Fourth Estate* more important far than they all." If he were speaking today I am sure he would enlarge the gallery considerably and provide ample space for the commentators and columnists who, obviously, know all about the world and its affairs, both scientific and otherwise. So much for the fourth estate.

The "fifth estate" of my title can best be described in the words of the distinguished scientist and technologist Arthur D. Little, who first used this term in an address in 1924 at the centenary celebration of the founding of the Franklin Institute.

> This fifth estate is composed of those having the simplicity to wonder, the ability to question, the power to generalize and the capacity to apply. It is, in short, the company of thinkers, workers, expounders and practitioners upon which the world is absolutely dependent for the preservation and advancement of that organized knowledge which we call "science."

The status of science and scientists in the 1960's is obviously a large subject, and here I will discuss four aspects of it which I feel should claim our attention, thought, and understanding. The first relates to the greatly expanded tempo, scope, and power evident in the development of science during the past quarter of a century. Secondly, I would like to consider the increasing role of science and technology as an instrument of national policy. A third area meriting attention is a changing pattern of scientific activities and some implications of this. Lastly, and most important for the future advancement of science, is the place of science and scientists in our modern social structure and the interactions with that structure.

Before elaborating on these topics, I think it desirable, even at the risk of covering ground familiar to many, to sketch briefly against their historical background some of the scientific developments familiar to us. In doing this I will attempt to emphasize not the content of science so much as the changing characteristics of scientific endeavor. For this purpose it will be convenient to have two reference points in time: the late 1800's just prior to the turn of the century and the decade from 1925 to 1935.

While the ranks of the fifth estate have grown rapidly in this century, it is still true that the number of scientists remains a small fraction of the total population. In the long years prior to 1900 the voice of science in national and world affairs was rarely heard, and the individual scientist working in his ivory-tower laboratory was a little-known member of society. Nevertheless, contributions of science and technology to human welfare and to the problems of the military, the growth of industry, and economic development in general were more important with each passing decade. Toward the end of the last century and in the early years of this one a new aspect of scientific activities began to emerge. This was the concept of highly organized team activity in scientific and industrial research. It was in Germany that this concept first appeared in any substantial measure, in the latter years of the 19th century. Its effective utilization gave Germany a leading position in producing such things as chemicals, pharmaceuticals, steel and machinery, and similar products of industries where scientific and technical knowledge was a prerequisite for effective production. This lead over other countries, including the United States, was retained up to World War I. Some of us can recall hearing the news in the early years of that war, before our entry, that the German submarine *Deutschland* had successfully eluded the British naval blockade and landed in Baltimore harbor. What may not be as well known is the fact that the cargo consisted of scarce pharmaceuticals and dyestuffs which sold at high prices in this country because of our almost total dependence on Germany for such synthetic chemicals. On the return voyage the cargo was mainly tungstic oxide, as tungsten was a critical raw material in many areas of Germany's advancing technology. Only after World War I and as late as the 1920's did the industrial research concept of today begin to appear as an important component of some of our own more technically based industries.

The world nitrogen supply and the fate of nations. Before this century, much scientific thinking was still limited in its scope and heavily circumscribed by the walls of the laboratory. There were of course exceptions. Though the ranks of science were small in number, they included a goodly share of giants—such men as Maxwell, Rayleigh, Herz, and Röntgen, to mention but a few. One in the field of chemistry was Sir William Crookes, president of the British Association for the Advancement of Science at its Bristol meeting in September 1898. In his presidential address, after an excellent analysis of factors bearing on world food supplies, he spoke as follows:

> The fixation of nitrogen is vital to the progress of civilized humanity. Other discoveries minister to our increased intellectual comfort, luxury, or else convenience; they serve to make life easier, to hasten the acquisition of wealth, or to save time, health or worry. The fixation of nitrogen is a question of the not too distant future. Unless we can class it among the certainties to come, the great Caucasian race will cease to be foremost in the world, and will be squeezed out of existence by the races to whom wheaten bread is not the staff of life.

That these are still matters of vital interest today is seen by recalling current discussions of the population explosion and discussions of this past fall relating to the sale of surplus wheat from this and other countries to help feed the millions in the Soviet bloc.

While we all have general awareness of the important role of organized industrial research in defense and in economic development, this can be focused more sharply by looking back at the events relating to the world's nitrogen supply that occurred after 1898. Based on fundamental research in Germany and Scandinavia, in the period between 1900 and World War I a new industry developed, that of nitrogen fixation. However, this was of only limited capacity at the beginning of the war. Since nitrogen is essential not only for agriculture but also for the manufacture of explosives, as war became imminent Germany began stockpiling Chilean nitrate. The first important naval engagement of World War I was fought not in the Atlantic but in the Pacific, off the coast of South America, in an operation in which British warships captured or sank a German merchant convoy carrying Chilean sodium nitrate back to Germany. With this event there were many predictions that Germany could not last long in the war, with her very limited domestic sources of nitrogen. As with many predictions, these proved quite wrong in the outcome. During the war years, the Germans, with their by then matured capability in industrial research, were able to build the first major nitrogen fixation industry in the world. Moreover, after her recovery from war and with the rebuilding of her commerce, Germany became the world's principal producer of nitrates and supplied these to Europe and the Atlantic seaboard at prices with which Chileans could not compete. The next step in this chain of economic events was a fiscal crisis in Chilean affairs, as a substantial part of the Chilean economy had been based for years on a tax on exported nitrates. Recovery from this crisis came only when, through research sponsored by American financial interests, more efficient ways were found of mining the Chilean nitre deposits and of extracting and marketing as a valuable by-product the significant amounts of iodine that they contained.

This illustration of changing conditions in the nitrate industry is but one of many that could be cited. In today's highly technological civilization, the fate of nations will depend increasingly on their store of scientific knowledge obtained through basic research and on their capacity and ingenuity in applying this knowledge to produce goods and provide services of all kinds. This is the basis of a sound economy and the key to its forward progress.

Rather than continuing with an account of more recent scientific and technologic events familiar to all, I will simply point out that greatly expanded basic and applied research between World Wars I and II and after World War II led to such results as the high state of development of the airplane for transportation, the whole electronics industry, the release of nuclear energy and its use for power and the propulsion of naval vessels, and, finally, the successful launching of orbiting satellites.

International Geophysical Year. More detailed review of similar developments would quickly reveal much to support the thesis that there has been a greatly expanded tempo, scope, and power in activities in science during the past quarter of a century. So far as scientists themselves are concerned, this could almost be regarded as the emergence of a kind of fourth dimension in scientific thinking. Justification for such a statement is evident in a number of directions—for example, in the thinking, planning, and execution that went into the project known as the International Geophysical Year. This was a bold frontal attack, involving international collaboration on a grand scale, which was made in an attempt to understand more fully the physical nature of the surface of our globe through a carefully planned survey of the scientific phenomena relating to the atmosphere, the oceans, and the input of radiation of all types to our near geosphere. The information gathered was vast, and the discoveries were many. Their significance for a better understanding of such important phenomena as weather changes and climatic cycles is already apparent. As the many scientists interested in this area continue working on the large number of data that were accumulated, a much deeper knowledge of the surface of our earth can be expected.

Another example of the type of thinking I have referred to, and one still on the scale of great dimensions, is the project currently under way which is known, for short, as "the Mohole." This is an attempt, fraught with great difficulty, to penetrate the earth's crusted layers to acquire a better understanding of the nature, composition, and behavior of its massive interior core. However, such scientific thinking and progress have not been confined to endeavors of large dimensions, even global in scale. In the past decade, work of a highly competent team of mathematicians, physicists, chemists, and biochemists at Cambridge University has led to a better comprehension of the basis of life processes, through discoveries of great significance in the field of molecular biology. The determination and unraveling of the complex molecular structure of giant molecules, such as ribonucleic acid (RNA) and deoxyribonucleic acid (DNA), have been major advances and outstanding illustrations of the effective collaborative scientific teamwork so characteristic of much current scientific activity.

Possible modification of the climatic cycle. A final example of thinking of this sort is a proposal by Ewing and others for possible modification of the age-old climatic cycle which results in repetitive glaciation of continental land masses south of the Arctic Circle. The geologic and related evidence from prehistoric times for the existence of such a cycle of ice ages with a period of perhaps 30,000 years appears clear. Ewing's thesis, in broad terms, is that the occurrence of this cycle is related to the extent of accumulation of ice and snow on the polar ice cap within the Arctic Circle and also the ingress and exit of warmer waters from the Pacific and the Atlantic over the edges of the fairly shallow geologic basin which holds the Arctic Ocean. The conclusion of the argument, which I shall not develop fully, is that this cycle could be altered by stopping or at least modifying the flow of water through the Bering Strait between the Arctic Ocean and the Gulf of Alaska in the North Pacific. This would indeed be a gargantuan project in applied science, execution of which could only have been thought possible—whether desirable is another question—with the availability of nuclear explosives. Thinking of this type would, in my judgment, have occurred but rarely in earlier periods of the development of science.

Much that I have outlined is evidence of the increasing role of science and technology as an instrument of national policy—the second topic under discussion. An illustration from behind the Iron Curtain at once comes to mind.

Few of us like the tenets of Soviet ideology, though many take complacent comfort in the disparity between the present standard of living in Russia and our own. Nevertheless, Russia has forged ahead through the encouragement of science, through the systematic employment of the methodology of research and development, and through the extension to large segments of her population of free education oriented strongly toward rigorous training in science and technology. Nicholas DeWitt, who has studied the Soviet manpower and educational system intensively, states the situation in the postscript to his book *Education and Professional Employment in the U.S.S.R.*:

> If the aim of education is to develop a creative intellect critical
> of society and its values, then Soviet higher education is an
> obvious failure. If its aim is to develop applied professional
> skills enabling the individual to perform specialized, functional
> tasks, the Soviet higher education is unquestionably a success,
> posing not only a temporary challenge, but a major threat in the
> long-run struggle between democracy and totalitarianism.

While DeWitt's first description of the aim of education may well give us pause when we think of values in relation to our own culture and society, and make us ask how well our own system of education has done, the validity of his concluding statement becomes apparent from the perspective of little more than a third of a century. In this short period Russia rose from the rank of a third-rate power to a position, today, second only to that of our own country.

National defense. There is a final element relating to the present role of science and technology as an instrument of national policy which must be mentioned. This concerns warfare and the preparation for warfare, or what is today euphemistically called national defense. From the first development of gunfire in the 14th century there had been little real innovation in the practice of warfare until this century, though the Civil War did bring the introduction of steel armor and the submarine. In World War I, gas warfare, tanks, and the aeroplane made their appearance. The development of the latter for military use between the two wars paced and enhanced the great development of commercial aviation, and this, in turn, has reduced passenger traffic on our widespread network of railroads to a fraction of its volume in the first third of the century. In World War II, radically new concepts such as the proximity fuse, the landing craft, and, of course, nuclear explosives were introduced. The war also saw the refurbishment and effective use of a very old device—the rocket. This was first used as a weapon by the Mongols about the middle of the 12th century, and it reappears from time to time in the subsequent history of warfare in various military versions. Francis Scott Key witnessed one of these occasions when he was a prisoner in a ship in the British fleet off Baltimore at the siege of Fort McHenry in 1814, and the spectacle inspired the line in our national anthem: "And the rocket's red glare, the bombs bursting in air. . . ."

The high state of effectiveness to which rockets were brought toward the end of World War II through intensive research and development and the advent of the German V-I's and V-II's provided the background for today's missile technology. Further development of long-range offensive missiles provided the launch rockets for orbiting satellites and for vehicles for space exploration. These are some of the advances that have completely changed the whole aspect of warfare in less than a third of a century. In this area there can be no doubt that scientific advance and capability are indispensable instruments of national policy.

What, then, have been the effects of this great expansion of science and technology, this changed scientific thinking, this involvement with national policy, on science itself, on its organizational patterns, and on scientists and their pursuit of scientific endeavor? The question brings me to my third topic. These effects have been both major in scope and diverse in direction. They have been both favorable and unfavorable for the sound advancement of scientific endeavor.

Consider first the positive side of the coin. Today, the nature and tempo of effective research requires ample funding for men, machines, and facilities, and funds have been made available in rapidly increasing measure during the past third of a century. A glance at one area, that of nuclear and high-energy physics, will quickly reveal the scale and pattern of support. From relatively small beginnings, in such laboratories as those of Rutherford and the Curies, nuclear physics in the 1920's and 1930's moved steadily but slowly ahead. Support for the first generations of high-energy machines, the early cyclotrons, came mainly from university funding and private giving by individuals or foundations. The demonstration, in the early years of World War II, of the feasibility of the nuclear chain reaction and of its significance for the release and utilization of nuclear energy in war and peace led quickly to federal support, first through the Manhattan District Project and later through the Atomic Energy Commission. The scale of this support was not in millions but in billions, and this pace continues today. However, the magnitude of expenditures, though indicative of the scale of modern scientific activity, is never a good measure of scientific achievement.

Astronomy. Nevertheless, a brief survey of several areas of science will reveal that great substantive progress has been made in recent years. A case in point is the field of astronomy. Next to mathematics, this is the oldest of the sciences, dating from Babylonian times in the 3rd century B.C., and it has a long history of achievement before 1900. The early years of this century saw the establishment, largely through private philanthropy, of a few observatories, such as that on Mt. Wilson, with telescopes and ancillary instrumentation larger and more effective, by an order of magnitude, than anything that had gone before. These were the forerunners of the large-scale scientific facilities familiar today—the giant cyclotrons, accelerators, and piles of nuclear physics. As astronomy moved ahead through the first half of the century, its progress was relatively slow by comparison with the burgeoning development of the laboratory sciences of chemistry and physics, which received much of their steadily increasing support from private, industrial, and government sources.

It was only after the establishment, at mid-century, of the National Science Foundation to support basic research that attention was turned to more adequate support for astronomy. In the middle 1950's the establishment by NSF of the Greenbank Observatory as a National Radio Astronomy Observatory marked a turning point in the character of federal support for basic science and fundamental research. From this beginning, federal funds became available for "national" basic scientific enterprises, such as the International Geophysical Year in 1957–1958 and, later, the Kitt Peak National Observatory in Arizona (near Tucson), the Mohole project, and the Atmospheric Sciences Center in Colorado. With this type of support, scientific progress in a number of relatively neglected fields, such as astronomy and various branches of the earth sciences, notably oceanography, was greatly accelerated, and the development of pure science for its own sake became, and now is, an acknowledged instrument of U.S. national policy.

The individual scientist and organized endeavor. It is of interest to consider the effects of these changes on scientists themselves. These are manifest in a number of directions, but here I mention only two, which appear to be the most significant. The first may be described as a type of dilemma with which the individual scientist appears to be increasingly confronted. In earlier periods the role of the individual scientist stood out clearly, and while the magnitude of his contribution might occasionally be large, usually it was small, though still real and discernible. Each small contribution was a piece in a growing mosaic of knowledge of the particular field involved. As this mosaic grew from initially few pieces of data and information, and as the basis for their interpretation and correlation became dimly recognized, there was ample scope for individual initiative, and there was wide freedom of choice and of action. As progress in the field increased, the few individuals with greater insight helped shape the pattern of the whole and made it part of scientific knowledge. Much of this was a seemingly random and quite haphazard process.

To this somewhat inadequate description of science in earlier years should be added the description given by Langley in his presidential address before the AAAS meeting in Cleveland in 1888. He characterized the pursuit of scientific research as "not wholly unlike a pack of hounds, which, in the long-run perhaps catches its game, but where, nevertheless, when at fault, each individual goes his own way, by scent, not by sight, some running back and some forward; where the louder-voiced bring many to follow them, nearly as often in a wrong path as in a right one; where the entire pack even has been known to move off bodily on a false scent. . . ."

Whether or not either of these descriptions is an adequate picture of earlier scientific endeavor, it is clear that, in spite of limitations of support, facilities, and equipment, there was ample room for individual freedom of choice and for the exercise of initiative, ingenuity, and resourcefulness. Out of this situation developed what we all inherit and cherish as the great tradition of freedom in science and of communication in science, both nationally and internationally. This may be stated otherwise by saying that science is universal and knows no bounds of geography, race, creed, or nationality. Many attributes characteristic of scientific endeavor in earlier periods still hold for the sharply quickened and greatly expanded domain of today's science. Unfortunately, there are signs that, as this domain grows further, as it becomes more highly organized, more pro-grammed, and more directed toward national and other ends, and as its impact on our culture and society becomes more widespread, some of this traditional freedom will be lost. An obvious example of this trend relates to freedom of exchange of information, so essential to the progress of science. In World War II it was found necessary to impose a cloak of secrecy and classification on research in the developing field of nuclear physics—research which led to the release and utilization of nuclear energy. All agreed that this secrecy was necessary in wartime, and it was imposed under the Manhattan District Project. In the early days of the activities of the Atomic Energy Commission these restrictions were still dominant, and it was only with the passage in 1954 of the "Atoms for Peace" modi-fication of the original Atomic Energy Act that some of them were removed or considerably relaxed.

This is one aspect of the so-called dilemma that many see ahead as the role of science becomes more important in modern civilization. Another, perhaps more important but more subtle, aspect can best be illustrated by an example from the field of chemistry. One of the great discoveries by Rayleigh and Ramsey at the end of the last century was that of the existence of the family of rare gases, the description of their properties, and the characterization of their chemical behavior. As these gases were studied further by many investigators, it became a tenet of chemical thinking that they were unreactive and would not combine with other elements and compounds. So strong was this belief that, as theoretical knowledge of chemical reaction and chemical bonding developed through this century, an essential element of each new theory of chemical bonding was that it should account for the supposed fact that these gases would not combine chemically with anything else. The first crack in this inviolate image came from the work of Bartlett, who demonstrated the combination of the rare gas xenon to form one of the components in a coordination compound of complex structure surrounding a central platinum atom. As often happens in science, the initial breakthrough was followed closely by others. Soon after Bartlett's discovery became known, further research and experimentation quickly destroyed this image that had dominated thinking in chemistry for some two-thirds of a century. The experiments leading to this final event need not be described in detail, but the circum-stances under which they were undertaken are relevant. While I cannot claim to know these circumstances at first hand, the account as it reached me, and as it is given here, is from a source I believe to be authoritative.

Under the system in AEC national laboratories which provides for research participation by scientists from outside the laboratory staff, a young physicist from a small college came to carry on research for a time at the Argonne National Laboratory. In discussing his proposed program with those responsible for general supervision of the Laboratory, he said he would like to attempt to react xenon and fluorine at an elevated temperature. Since most physical scientists were convinced that the rare gases were unreactive, and since this reaction had already been tried in the Argonne Laboratory at ordinary temperatures, it is reasonable to assume that in the discussion that ensued doubts were raised about the wisdom of devoting the investigator's time and the resources of the Laboratory to the attempt. If such doubts were raised, at least they did not prevail, and it was agreed that the young physicist should go ahead with the attempt. The result was a spectacular, unanticipated discovery in the field of chemistry.

When a mixture of xenon and fluorine was heated in a nickel container to 400°C and then cooled rapidly to room temperature, a deposit of white colorless crystals of the compound xenon tetrafluoride was found, and the long-standing belief that rare gases are inert was shown to be a myth. These events came to their culmination in the summer of 1962. On learning of this discovery, many investigators at the Argonne Laboratory and elsewhere went quickly to work and made other compounds of xenon and fluorine, as well as of certain of the other rare gases. As of the end of 1963, there is already an extensive literature relating to such compounds. In passing, and somewhat out of context, I might note that *Science*, in its "Reports" section [**138**, 136 (1962)] carried the first general news of this important discovery through a communication from the Argonne group dated 2 October 1962. Incidentally, the interval of 10 days from 2 October to 12 October, the date of the issue in which the report appeared (which carried a striking picture of the crystals of xenon tetrafluoride on the cover), probably constitutes an all-time record in the rapid communication of new scientific information through the printed word.

More in the context of the present discussion of the environment in which today's scientists work was a very timely and thoughtful editorial in the same issue by the editor of *Science*, Philip Abelson, entitled "The need for skepticism." The last paragraph of this is well worth quoting.

> There is a sobering lesson here, as well as an exciting prospect.
> For perhaps 15 years, at least a million scientists all over the
> world have been blind to a potential opportunity to make this
> important discovery. All that was required to overthrow a
> respectable and entrenched dogma was a few hours of effort
> and a germ of skepticism. Our intuition tells us that this is just
> one of countless opportunities in all areas of inquiry. The
> imaginative and original mind need not be overawed by the

imposing body of present knowledge or by the complex and
costly paraphernalia which today surround much of scientific
activity. The great shortage in science now is not opportunity,
manpower, money, or laboratory space. What is really needed
is more of that healthy skepticism which generates the key
idea—the liberating concept.

Of serious concern under present conditions of highly organized and
programmed scientific endeavor is whether the freedom, initiative, and
originality of the individual will still be able to emerge to play their important roles, so evident in the history of science in earlier periods. It is
disquieting to speculate on what the ultimate outcome would have been,
in the case cited, if it had been decided not to make the experiment. How
long would it have been before the proper conjunction of circumstances
occurred again—the individual with faith in his idea and skepticism of
established dogma; a laboratory with chemists experienced in handling
potentially dangerous fluorine reactions; and last but not least, a super-
visory group willing to authorize the trial? Here the conjunction of events
was propitious, and the outcome was a brilliant success. Unfortunately,
or perhaps fortunately for scientific morale, as science progresses the
number of instances in which the circumstances are not propitious is
unknown. We can only hope it is small.

Specialization. A subject of much current interest is the rapidly increasing degree of specialization in science, which has paralleled science's
growth and expansion in the past 35 years. Consideration of this is
important, because of its implications for sound scientific education and
also because of the common reaction of the lay public to highly specialized
activity of any sort. Specialization in the most general sense is not new.
However, when we consider the complexity of our own social structure—
the profusion of implements, machines, instruments, and devices—and its
specialisms of all kinds, we tend to think the latter are characteristic of,
and even in a measure unique in, our society and time. A moment's
reflection will indicate that such is not the case. The thoughtful citizen
of the great ancient metropolis of Rome, with a population of nearly 2
million persons in the 2nd century A.D., must have been confronted with
something of the same situation. The highly organized civilization of the
Roman Empire must have required a high degree of specialization on the
part of its citizens to provide its food supply, build its aqueducts and
public works, and maintain its roads and the government of its far-flung
provinces and colonies—not to mention the high state of development of
literature and the fine arts. It seems clear that elaborate specialization,
comparable in scope to our own, has been a characteristic of all great
civilizations, especially those which were highly urbanized.

Nevertheless, it is desirable to consider briefly the nature of specialization itself, particularly that in the realm of intellectual endeavor. Here the
intense concentration of an individual on a limited area of special knowledge
and his attainment of expertness in his field tend to break the broad pattern

of uniformity of the social structure. This is especially true in a democracy. The resulting separation of the individual from the stream of the common affairs of man tends to make the average citizen uneasy. Shaw put this feeling succinctly, in *The Doctor's Dilemma*, when he said, "All professions are conspiracies against the laity."

Much of the extensive specialization in the sciences has some features that can be most clearly delineated by the following comparison: "A salesman is one who begins by knowing a little about everything and who goes on learning less and less about more and more until he ends up knowing practically nothing about everything." On the other hand, "A specialist is one who starts off knowing a great deal about very little and goes on learning more and more about less and less until he ends up knowing practically everything about nothing."

For our present purpose the description of the salesman can be ignored. That of the specialist will bear further scrutiny. The difficulty lies in the common lay conception that what is small or restricted in scope and dimensions is simple, and in its limits amounts to "nothing." Here is one clue, and a very significant one, not only to the common negative reaction to specialization in general but to the general public's understanding of specialization in science.

Scientists, unlike the lay public, have the privilege of appreciating the accomplishments of a truly great specialist as he reveals fascinating glimpses of things to come, when, from time to time, there is a breach in the ramparts that bar us from comprehension of nature. These ramparts are long and formidably complex, as Vannevar Bush implied in his description of science as "the endless frontier." Rarely do they succumb to attack along a broad front; when they do, it is only through the work of a genius—and geniuses are rare in the human race. If science is to move forward, it will be increasingly important that the general public acquire a better understanding and some appreciation of the true nature of scientific specialization.

As we look ahead to yet unconquered areas, we may confidently predict that soundly conceived specialization in science will continue to survive and multiply. Historically, much of the effort in science has related to inanimate things, or to relatively simpler organisms or functions. As our growing knowledge permits us to move more firmly to studies of human behavior and of its psychological, biochemical, physiological, genetic, and other bases, it is possible to envisage new coalitions between psychologists, neuroanatomists, and neurophysiologists; as they grow, these coalitions may develop as specialties, as is the case for present-day biochemistry and biophysics.

With this prospect confronting us, we shall have to consider the negative aspects of the further growth of specialization and to constantly appraise its soundness. This will be especially desirable in developing sound principles to be followed in future education in the sciences. Here the danger is that the form may be mistaken for the substance. To illustrate the problem and not invoke invidious comparison, let us imagine some future specialty that we call neurobehaviorism, for want of a better designation. On what will the validity and worth of such a field, both as a contributor to our knowledge and as a field of endeavor, depend? First, it will depend on how well those in the field are versed in fundamentals of the relevant derivative sciences, such as neuroanatomy, neurophysiology, and neurobiochemistry. Beyond this, and of great importance, it will depend on how well they understand, or can acquire understanding of, principles from the underlying basic disciplines of psychology, mathematics, physics, chemistry, biology, and physiology that are relevant and applicable to the field in question.

By this criterion, the validity and worth of an area of specialization would depend on the firmness and clearness of the pathways from the outer branch to the deep, sound roots of available scientific knowledge. Perry relates an episode that occurred at Harvard in the 1830's, about Ralph Waldo Emerson and Henry Thoreau, which has point in the present context. The then-young naturalist, who was an intimate of the Emerson household, sat quietly in a corner one day while Emerson expounded to English visitors on education at Harvard, saying, "At Harvard College they teach all branches of learning." At this point Thoreau, to the embarrassment of his patron, blurted out, "Yes, but none of the roots." Without vital and continuing sustenance from strong roots, the branchces of specialism will bear meager fruit.

Let us now turn to the fourth topic of this discussion of the fifth estate in the 1960's—the impacts of these changes in science on our current culture and the response and reaction of the latter to the change. I have already mentioned many of these changes and need not review them, but two additional ones deserve attention. However, before considering these let us look at a few figures for the sake of perspective.

Scientists and technologists have been, and still are, a relatively small minority group in our total population. In 1900 they numbered perhaps 90,000, representing little more than 0.1 percent of a population of about 76 million. Federal expenditures for science in 1900, similar to the federal expenditures for research and development of today, were about $10 million, or between 0.5 and 1 percent of the annual federal budget. The corresponding rough figures for 1963 are, 2.7 million scientists in a population of 190 million and federal R&D expenditures of $14 billion, which now require about 15 percent of an annual federal budget of the order of $95 billion. Thus scientists, though their number has increased 30-fold since 1900, still comprise a relatively small part, about 1.4 percent, of the total population.

Effects of the drain on federal resources. It is important, first, to consider some of the consequences of this increasing drain on federal resources that is caused by the recent, almost exponential growth of science and technology. Since federal revenues grow at a much slower rate than the economy does as the economy advances, it is obvious that some adjustment in the growth rate of federal expenditures for science must take place. Indeed, this is already occurring, as is evident to anyone who has followed recent hearings before Congress relating to the expenditures for science projected for the next annual federal budget.

On aspect of this adjustment poses a new and serious type of problem that scientists have not faced previously in any substantial measure. With limitation necessary, on what principles is the assignment of priorities to projects in the various fields of science to be made? What are the relative merits, both in a scientific sense and from the standpoint of the national interest, of a new, large accelerator for nuclear physics, costing perhaps $100 million for its initial construction and about a third of that amount for its annual operation; of the Mohole project for drilling through the earth's crust, variously estimated to cost between $50 million and $100 million; of the expenditure of similar sums annually for biomedical research on cancer or the diseases of the heart; and of landing a man on the moon by 1970, at an estimated cost of over $5 billion? There are no clear guidelines on which to base such priority decisions, and their formulation will require a higher order of statesmanship among scientists and those in the upper echelons of government than has existed heretofore.

Effects on health. The rapid growth of science has had a second type of impact on society in our greatly expanded technological and industrial civilization. This growth has been so great that it has already begun to alter man's traditional natural environment. The emerging problems involve such things as air and water pollution, radiation hazards, occupational hazards, and contamination of milk and food supplies, which are now classed under the general head of environmental health problems.

Several years ago I was asked by the Surgeon General to head a committee of 24 members from widely diverse scientific disciplines. The group was to analyze and survey the problems in the environmental health area and to make a 10-year projection of the nation's needs for scientific research relating to environmental health and of its needs for trained manpower to deal with the problems.

These problems are varied, complex, and serious. They range from the provision of adequate sewage disposal for large and growing metropolitan districts to the recently noted higher level of radioactive contamination of caribou meat, which is an important part of the diet of Eskimos in northern Alaska.

The origin of the high levels of radioactivity in caribou meat was relatively simple to trace and understand, though not necessarily easy to control. Certain lichens on which caribou feed were found to absorb relatively larger amounts than most plants of the radioactive trace elements which the soil had received from the debris of fallout.

A simple illustration relating to the matter of sewage disposal in large metropolitan areas will show the complexity of many of the problems involved in environmental health. A number of years ago, in order to handle its sewage disposal without contaminating Lake Michigan, Chicago built a drainage canal in which water from the lake ran across country to empty into the Mississippi. The sewage effluent from Chicago was fed into this artificial running stream. The dilution of the effluent by water from Lake Michigan reduced its concentration to the point where the organic sewage could be oxidized effectively by the dissolved oxygen in the canal waters, and well-purified water was delivered to the Mississippi. As the city expanded industrially, steam plants were built along the canal, and these discharged warm water from their condensers into the stream. Ultimately, the effect of these additions of warmer water was sufficient to raise the average temperature of the canal water by some few degrees throughout the year. With this development the phenomenon technically known as "heat pollution" became operative. Simply stated, the higher temperature reduced the concentration of oxygen in the water, and therefore the capacity of the flowing stream to oxidize the organic matter present. As of several years ago this "heat pollution" had reached such proportions that its effect on the sewage disposal problem of the Chicago area was estimated to be equivalent to the effect of adding a million people to that metropolitan area.

These are but a few examples of the many effects on the economy, on health, and on various aspects of our culture and society of the greatly increased endeavors of scientists. What has been the reaction to these great changes occurring in little more than a quarter of a century? Here two things are relevant—the status of the general public's knowledge of science and its methods and, even more important, the image in the public mind of the whole modern scientific enterprise. A realistic appraisal of these two factors does not give much ground for thinking that the public has a sound comprehension of science.

The nonscientist's view of science. For the great majority of our people, formal education terminates with high school. Sober reflection about our educational system, after the launching of Sputnik, clearly revealed the woeful inadequacy of the science education of most of our people as a basis on which to build any real understanding of modern science. Since Sputnik, real improvement has been made in science teaching in many of our lower schools, but the effects of this in the adult population will not be evident for another generation.

Given this lack of any sound comprehension of science, what picture can be drawn of the image of science in the public mind? This image is difficult to describe, for it is compounded of many diverse elements. These include respect and gratitude for the "miracles" of modern medicine; admiration for the know-how of the applied science which can put satellites in predetermined orbits; and awe, verging on fear, of the results of the mysterious release of nuclear energy. Two events in our time must have contributed greatly to the building of such an image, since the public, as well as most scientists, had no warning and little preparation for their advent. The first was President Truman's unheralded announcement of the dropping of the atomic bomb—a spectacular but terrible demonstration of the power of modern science. The second was the sudden news, one day in October 1957, of a second satellite orbiting our planet. This was Sputnik, the first of a growing family which later included Echo I—a "star" whose rapid course across the night sky could be easily followed with the naked eye.

What is the significance of this image, and of this lack of real understanding by the general public, for the future of science and scientists? Some already feel that scientific endeavor must be controlled and circumscribed if it results in pollution of air and water, in contamination of food with pesticide residues, in the hazards of radiation and the development of nuclear weapons. Still others, left ever farther behind in their understanding of rapid scientific advance, take refuge in a polite, but neutral, type of anti-intellectualism toward all scientific activity. The emergence of attitudes like this among nonscientists is the basis of Snow's discussion of the two cultures, and of his warning that the rift may grow wider unless the trend is checked.

Faced with these possibilities, what should we, as scientists, do? We are in some sense a privileged minority group, and all of us should be ready to exercise the grave responsibility which we all share, "to increase public understanding and appreciation of the importance and promise of the methods of science in human progress." These words are quoted from a statement of the objectives of this Association. A second objective of our organization is "to improve the effectiveness of science in the promotion of human welfare." These two should be the articles of our scientific creed in the years ahead. Furthermore, as scientists, we should not lose our perspective but should recall the history of science and remember that it has survived pestilence, wars, and disaster and has surmounted barriers of race, religion, and language. Beyond this, it is even more important to recall, in a gray period of international tension, that all members of the human race, throughout its evolution and long history, have had a common opponent. This is inscrutable nature with her seemingly inexorable laws, her hosts of organisms and parasites, her hurricanes and catastrophic events of all kinds. For our human race the central problem is still that of understanding nature and attempting to control it. Here the thinking and tools of modern science have a great contribution to make. May we use them well.

Much of what I have said of warnings, of impacts and reactions, and of grave concern may have the ring of pessimism for the future as science moves swiftly ahead in one of the great adventures of the human mind. That this is not my intent can be made clear by a closing quotation from Carlyle's great satire *Sartor Resartus*. In this he attributes to his fictitious author, "Professor Teufelsdröckh of Weissnichtwo," these words, in the promethean spirit of which I share: "Man's unhappiness, as I construe, comes of his Greatness: it is because there is an Infinite in him, which with all his cunning he cannot quite bury under the Finite."

The Changing Environment of Science

Alan T. Waterman
Montreal, 1964

Alan Tower Waterman (1892–1967), *a physicist by training, spent half of his professional career in science administration, including 12 years as the first director of the National Science Foundation. Waterman completed undergraduate and graduate work in physics at Princeton University (B.A., 1913; M.A., 1914; Ph.D., 1916) and taught physics at the University of Cincinnati until joining the Army Signal Corps in 1917. From 1919 until World War II, he was professor of physics at Yale University. Waterman's research during this period focused on electrical conductivity, photo-electric and thermionic emission, and achievement testing in physics. In 1942 he joined Karl Compton as vice-chairman of an Office of Scientific Research and Development division which was responsible for military instrumentation, such as radar. Later in the war effort, he was Compton's deputy, and then chief, of the Office of Field Service.*

During the lengthy postwar period in which Congress struggled with the complex issues of government-science relations, Waterman took charge as chief scientist of the Office of Naval Research. The National Science Foundation, created in 1950, grew at an unprecedented rate during Waterman's tenure as director; this growth is reflected in annual outlays, which jumped from $10 million to $300 million. After his retirement in 1963, Waterman continued to serve as advisor, board member, and consultant to numerous public and private agencies. He was a fellow of the AAAS (president in 1963) and received honorary degrees, national scientific awards, and the Presidential Award of Freedom for his leadership in support of scientific research.

I suppose it is fair to say that scientists as a class have deeper concern about the present state of the world than most groups. This concern is natural and understandable. It undoubtedly had its origin in World War II in such dramatic developments as the atomic bomb and biological warfare, which disclosed new and awesome possibilities of man's destroying himself through the findings of scientific research. Since that time the picture has broadened and changed considerably. With the cessation of active warfare, the contribution of scientists to the development and use of military weapons and devices has lost much of its compulsive quality and has returned to a more normal state.

However, the change of greatest import to scientists developed as an outgrowth of the cold war. It is the general realization that the entire future of a country, not just its military might but its economic strength and welfare, depend markedly upon its progress in science and technology. This has brought scientists into prominence as the potential saviors of their countries, a most embarrassing position for any group but especially for ours. In the past, we have tried to avoid publicity; it is a disturbance to calm and concentrated thinking. Rightly or wrongly, approbation of the public has been of relatively little importance to us; it is the recognition and respect of our colleagues that counts. In normal times our allegiance is strongly to our science; to attempt to direct our efforts toward causes of national importance is ordinarily confusing and disturbing. We have a fundamental conviction that the country's cause is best served when we are given *carte blanche* to work as we see fit, since we know very well that the greatest progress in science is made when that is the case. But this is an oversimplification. We must admit that the demands of technology have been present from the very beginning. Archimedes produced his engines of war, Galileo studied the operation of well pumps. Newton gave serious attention to navigation. Jenner had his cow pox problem and Pasteur his beer project and the prevention of infection. And, as a matter of fact, such pressures have been responsible for much important progress in science itself. For some time now the feedback from technological innovations, themselves stimulated by basic research, has made countless techniques and instruments available for fundamental research.

Thus, in an important sense the conflict between science and technology is not in itself a real conflict of interest. The conflict occurs principally because of the competition between the two for money and manpower. The public, unable to distinguish clearly between them, gets into the act by insisting upon prudent, economical, and understandable use of its (public) funds—that is, tangible and useful results. Unfortunately for many academic scientists, life has a way of presenting insistent practical problems whose solutions require their attention.

Within science itself, however, one finds an increased sense of responsibility for our future. This is not due solely to the strengthening of the popular image of the scientist; it is heightened within science itself—in many areas which give high promise of progress, notably in biochemistry and genetics; in nuclear physics, with its potential for power from nuclear fusion; in exploitation of the oceans for food and minerals; in attempts at weather modification, and in the exploration of outer space. This promise is augmented by the extension of knowledge to such incalculably remote domains as the atomic nucleus and galaxies nearly at the boundary of the known universe, not to mention the almost unbelievably complicated structure and behavior of the living cell.

This sense of responsibility is spurred by the reputation scientists have acquired with the general public, which has served both as a stimulant and a sobering influence. On the one hand every country has come to believe that its salvation lies in technology—usually misnamed science. The technical industries have looked to their research departments and their research analysts for the most dependable forecasting of profitable lines for future development. The man on the street seems to view all this with mingled feelings. On the whole he is grateful for progress in health measures, communication, housing, transportation, and the many technical conveniences which both simplify and complicate his existence, despite occasional feelings of resentment over the increasing novelty and complexity that surround him. When it comes to major events, such as possible devastating nuclear war, fallout, or the costly but intriguing conquest of space, the element of anxiety is added. "Why isn't there some way," he asks, "to keep our technological advances within safe and prudent channels? Why must these troublesome questions arise? Can't scientists be persuaded to work only constructively? If they can't do that, why don't we limit their activities by providing money only for selected desirable and noncontroversial enterprises?" Indeed, some would go so far as to advocate a moratorium on research in the natural sciences, on the one hand to avoid such disagreeable issues as those posed by nuclear and biological warfare and, on the other, so they say, to allow the social sciences to catch up and solve these vexing problems before the natural sciences make them too tough.

In the meantime, how has modern society been dealing with all this? What degree of attention are nations giving to this subject? To what extent do their governments participate in the conduct or support of scientific research and development? What sort of policies are emerging? These are questions which the Organization for Economic Cooperation and Development (the OECD) has canvassed among its member countries, a task for which we owe it a debt of gratitude.

In the OECD observations regarding general aspects of science policy, the following points were apparent. In the first place, education is a critical factor, since it must provide the human resources for technological progress and because it creates a favorable psychological climate. Next there must be up-to-date provision for the numbers and skills required in the labor force. An important consideration is the training of potential research workers, and especially of future managers. It appears to be generally agreed that scientists and engineers of high capability are desirable in management positions, in both industry and government. Finally comes the training of the research scientists and engineers themselves. This necessitates high quality in the graduate and postgraduate facilities at universities and institutes of technology.

Immediate potential for meeting these criteria of course exists in a number of countries. In others, further development is required, and in general this becomes a responsibility of the respective governments.

All countries face the problem of securing a satisfactory output of trained scientists and engineers. In most countries this problem is caused by a very rapid rise in research and development expenditures over the past decade; this rise greatly exceeds the increase in the gross national product. Generally speaking, the ratio of R&D to the per capita GNP is high (1 to 2 percent) in the large industrial countries; in these, industry performs two-thirds or more of the R&D. In countries where there is strong emphasis upon agriculture, forestry, mining, and fishery—such as Australia, Finland, Canada, and Norway—the ratio of R&D to the per capita GNP is lower and industry performs only about one-third of the R&D. In almost all countries the government finances most of the R&D.

In practically all countries fundamental research is primarily conducted at universities and nonprofit institutions and is increasingly receiving support from government. In the United States, the United Kingdom, and France, the government finances applied research and development largely through contracts with industry. Such financing occurs to a much lesser extent in Canada, the Netherlands, and Japan. In Canada, I understand, nearly half the total R&D is performed in government laboratories.

Whereas earlier, in most countries, provision for R&D funds was chiefly the responsibility of the Ministry of Finance or the equivalent office, the more advanced countries are giving increasing attention to the formation of top advisory councils to their governments, which work in cooperation with the finance office. Likewise, in the more advanced countries the influence of the national academies of science has grown, in providing advice to government and in setting the national and international tone for scientific achievement.

Such is the situation facing us and other nations at this stage of development, and such are some of the ways in which we and they have moved to meet our problems.

Because of the mounting commitments to science and technology, much talk and some concentrated thought and study have been directed toward improving the effectiveness of our efforts through planning, management, and education. There have also been attempts to evaluate the impact of the national effort upon our economy and our national achievements. These are important questions, and it is earnestly to be hoped that the current efforts will stimulate concerted and continuous study among economists, social and natural scientists, educators, and administrators. The issues are complex and are not likely to be solved by *ad hoc* committees or conferences alone.

Insofar as this activity sharpens the focus of our attention upon the identification and definition of worthy objectives, their relative priorities, and the feasibility of proposed means of achieving them, it must be regarded as very worthwhile. However, two important caveats should be heeded:

1) The planning for the identification and pursuit of technological objectives, no matter how feasible or worthy, should not be permitted to monopolize the national effort at the expense of science, and of basic research in particular. Such a policy leads in the long run to diminishing returns and ultimate stagnation.

2) Any attempt to forecast detailed money and manpower requirements for free research in the component scientific disciplines is, in my opinion, a questionable undertaking, no matter how experienced and distinguished the reviewing body. Applied research will always receive this kind of attention. But such attempts for free research introduce a concerted extrapolational bias into the system and sound an authoritarian note. Besides, what stronger motivation can there be for creative, original research than the individual scientist's own evaluation and decision as to the most promising course for him to pursue? As history abundantly proves, the capital discoveries in science generally lie in the unknown and cannot be predicted or planned for—and these may occur in any branch of science.

I shall not pursue further this topic of the impact of science and technology on society. Rather, my purpose is to invite your attention to the effect of this radical and sweeping transformation of activity upon the progress of science itself, stressing science in its traditional sense of the "search for truth." In the dynamical center of this interpretation lies basic research—the systematic and specialized search for knowledge and understanding. But of course science is more than this; it is the organized and classified body of knowledge which results from the search. Research is merely its frontier.

With the recent universal recognition that science and technology are essential to the progress of civilization, and with the attendant glamor which attaches to research, the environment for science has altered. For most of its history the devotees of science have been attracted to its study not primarily for the purpose of securing information that might be useful in some practical way but, like Kipling's elephant's child, out of "satiable curiosity," in the search for new knowledge no matter where or what it might be. With this motivation dominant, the search for knowledge in science has proceeded without boundary or limit. Scientific exploration mushroomed out in all directions, encompassing a range which would have been impossible under concerted planning. Of course many extremely important advances occurred by reason of some practical need or incentive, but by and large the scope and range of scientific investigation was not dictated by such considerations.

During the present century the technical industries, whose existence depends upon successful practical development and production, have increasingly come to conduct research themselves. Many of them have even recognized the advantages of pursuing basic research in areas where such work will lead to better understanding of their technological problems. This trend was accelerated during and after the war by such sensational results of research and development as atomic energy, radar, and the transistor. Nowadays no progressive technical industry or government bureau would attempt a large developmental enterprise without careful survey of the underlying research and, where necessary, inclusion of such research as part of the developmental process.

However, a larger proportion of support is provided for applied research than for basic research; in the U.S. the ratio is about 2 to 1. There is a corresponding majority of scientists employed by industry and government as compared to academic and other nonprofit institutions. For engineers, the ratio is of course much higher than it is for scientists..

But this is not all. There has been a steady increase in the support of basic research which may be termed "mission-related"—that is, which is aimed at helping to solve some practical problem. Such research is distinguished from applied research in that the investigator is not asked or expected to look for a finding of practical importance; he still is exploring the unknown by any route he may choose. But it differs from "free" basic research in that the supporting agency does have the motive of utility, in the hope that the results will further the agency's practical mission. A considerable body of basic research is receiving support because it is so oriented. Thus, basic research activity may be subdivided into "free" research undertaken solely for its scientific promise, and "mission-related" basic research supported primarily because its results are expected to have immediate and foreseen practical usefulness. Much of the emphasis upon basic research in the areas of cancer and solid-state physics illustrates "mission-related" characteristics. Since the support of "mission-related" research is easier to justify, when budgets become tight it tends to survive at the expense of "free" research. This tendency, when coupled with the present preponderance of "mission-related" research support, could prove a serious detriment to the progress of science, by curtailing free research and by concentrating too much effort on trying to solve practical problems that currently appear insoluble. As Oppenheimer has pointed out regarding progress in research, "in the end you will be guided not by what it would be practically helpful to learn, but by what it is possible to learn."

Some idea of the relative magnitudes of national funds provided for "mission-related" and for "free"' research, respectively, may be obtained as follows. Let us assume that all funds available in the following categories are provided for "mission-related" basic research: basic research by industry, by government laboratories, and by academic institutions from grants or contracts received from government agencies having practical missions. The latter chiefly include the Department of Defense, the Atomic Energy Commission, the Aeronautics and Space Administration, and the Departments of Health, Education, and Welfare, Agriculture, Commerce, and Interior—agencies authorized and encouraged to conduct and support basic research related to their missions. On these assumptions, about 80 percent of the total national funds for basic research are provided for the "mission-related" variety, and only 20 percent for "free" basic research. These figures are far from precise, of course; the assumptions are oversimplified, and many agencies are liberal in their interpretation of what basic research may be useful to them. But the approximate magnitudes of the figures are significant, and illustrate my point.

Two observations concerning this are in order. First, if industry were to confine the research activities of its laboratories strictly to applied research and if the government were to place similar restrictions on agencies with practical missions, leaving the support of basic research entirely to a single federal agency, to private foundations, and to universities, it is reasonably certain that the support of basic research would drop to a mere fraction of its present figure. Second, while this might be attractive in budget circles, such a course would be disastrous not only to science but also to technology.

In raising the question as to the extent to which basic research is supported for essentially practical reasons, I wish to be entirely clear on one point. It is not my purpose to question the importance and desirability of applied research or of basic research which is intended to provide better insights into developmental applications. Both are highly desirable and necessary, and science should play a direct part in their encouragement. They appear to be logical steps in the march of civilization in that they represent progress in providing for necessities such as food, housing, health, communication, transportation, and the national defense. They also contribute attractive innovations in our way of life—comforts, pleasures, opportunities for using leisure, and freedom from routine and drudgery. Of longer-range significance are their effects upon control of our environment; upon extension, in magnitude and in kind, of our sources of commercial power; and upon the discovery and exploitation of natural resources. Above all, whether we like these developments or not, of one thing we may be certain: the forward march of technology is inevitable. This is an important lesson of history, from the discovery of fire and the invention of the wheel and the lever onward. It is a lesson which has withstood the ravages of heat and cold, famine and pestilence, and many ideological conflicts. The convincing proof of this doctrine is contained in science itself—the science of evolution—as the powerful contribution of technology toward survival, and indeed toward increasing domination over environment. During the present century, we are witnessing perhaps the greatest triumph of this doctrine in the conversion of all mankind to acceptance of the thesis that science and technology are essential to survival. This appears to be a thesis to which one must subscribe if one believes in the progress of civilization as we know it.

But is this the whole story? From time immemorial man has evolved religions and philosophies representing his conviction that there is more to life than merely its physical aspects. Through imagination, study, and inspiration he has put forward philosophies, modes of conduct, and ways of life that concern the motivations and the aims of the individual and of society. From quite early times science has been thoroughly involved in much of this thinking, as evidenced by the earlier designation of natural science as "natural philosophy." From the record it is clear that, even after this term had fallen into disuse, science continued to have profound influence on philosophical thinking. It still does; witness the number of distinguished scientists of the present century who have written authoritative works on the subject—men such as Whitehead, Eddington, Jeans, Bridgman, and Dubos.

To what extent does this motivation for science still exist? How important is it? Are we observing, or failing to note, the gradual development of a monopoly by research oriented toward practical ends? There will of course always be individuals who firmly believe in the independence of research activity and who strongly wish to carry it on in the traditional academic manner. Will this group diminish in numbers or become frustrated? At the same time there appears to be a rapidly growing body of scientists employed in industry and government whose motivations are mixed, who believe in the support of basic research of the free variety but feel that "mission-related" basic research should have a higher priority, and still others who believe that research should be justified entirely on the basis of its specific utility.

Any uncertainty as to the importance of this question should be dispelled by looking into the history of science and noting: (i) the impressive discoveries made solely in the interest of pure science, and (ii) the statistical evidence that most of the body of science ultimately achieves practical utility.

Thus, even if we admit the requirement of utility as the prime justification for basic research, we still must allow free research to be included. It must be concluded that, in the long run, practical accomplishment will be greatest if in the support of basic research there is no limitation of the research to areas of foreseen practical importance.

I am reluctant to leave this topic without mentioning the thesis to which many, including myself, subscribe, to the effect that completely free research is highly important in its own right, not solely because of the probability that it will progress more rapidly and ultimately produce practical and tangible benefits, but because of its stimulating effect on the imagination and its philosophical implications concerning the universe and man's place in it. Who can say that ultimately this may not be the most important consideration of all?

I wish now to consider a different but related feature of my topic: What is the secret of the power and influence of science in the most fundamental sense? Is its source of strength at all in jeopardy? If so, are there any steps we ought to take to safeguard its future?

This body of knowledge—science in the modern sense—has steadily developed over the past 400-odd years into a most imposing edifice. Once science had discovered the art of experimentation it found a way to test hypotheses and speculation regarding the nature and behavior of the physical world and thus established a powerful base for drawing objective conclusions. This, together with the development, along with mathematics and logic, of techniques of classification and analysis, united the findings of science into a structure of extraordinary strength and stability. Furthermore, this technique has had a highly democratic flavor: anyone can challenge the alleged facts and theories of science. If he can prove his point within the scientific community by observations, experiments or reasoning that others can repeat and verify, then his contribution becomes an integral part of the body of science. Science has thus acquired a respect and confidence on the part of literate mankind that is unique. In consequence, the findings of science have a logical validity which is unmatched in other fields of human thought. At the same time, in a most interesting manner science remains flexible, since important new findings may necessitate revision of existing points of view. Generally speaking, and contrary to popular view, these revisions commonly take the form of refinements or increased generality and only occasionally bring about a revolutionary overthrow of existing principles. The impressive result is that the edifice of science has a strength and stability which is dynamic and resilient rather than static and brittle.

How do we account for these characteristics? They appear to be due to the maintenance of a broad base of inquiry; to the exercise of a lively imagination; to the utmost objectivity in search and logic; to a sense of proportion and urgency in the selection of scientific objectives. One must also recognize the necessity of built-in mechanisms for coordination, cross-fertilization, and collaboration, and finally—most important of all—of a creative dedication. These are high ideals, not commonly encountered or possible to the same degree in most other areas of human affairs and requiring a high degree of motivation and integrity.

These principles and this code of behavior are thoroughly learned by every researcher, beginning with his years of graduate study. It has been a source of the greatest strength to the body of science that, on the whole, these principles have been scrupulously observed. There has been no means of enforcement other than public opinion within the scientific community. Just as the standing of an individual in his field of research rests primarily with his colleagues, so too does his reputation in his behavior as a scientist. The real strength of this philosophy lies in the fact that these principles are essential for sound progress in science.

Thus, much of the power and stability of science has rested upon the sense of dedication and integrity of the community of scientists. Not only has this been thoroughly incorporated in their indoctrination but it has been further developed and fostered as a code of honor among scientists: to be scrupulously objective in their research, in their reporting, and in giving credit where credit is due. It would seem that the chief reason for the almost universal observance of this code has been that the scientist desires the respect and confidence of his colleagues, rather than recognition before any other audience. Anyone departing from these rules of behavior is ostracized by his kind.

In most careers, however, loyalties and motivations are more complicated. They involve such considerations as allegiance to, and recognition by, one's employer and his organization, one's community, church, political party, and friends, and the public generally. An interesting question is the extent to which these other loyalties are increasing in importance among scientists and encroaching upon loyalties to the scientific community. If so, will this warp or weaken the edifice of science or retard its progress?

Profit institutions such as industrial laboratories are, of course, clear examples of organizations that require strong loyalties in carrying out purposes related to the well-being of the organization. The same may be said of government establishments. Because of the increasing proportion, in the scientific community, of scientists employed by industry and government, these considerations are inevitably coming to receive more and more emphasis.

Likewise, with increasing dependence of colleges and universities upon the federal government, federal support of scientific research at these institutions becomes more and more strongly related to their health and strength. Again, this may be manifested in tangible or intangible pressures on the part of academic institutions for their scientists to engage in sponsored activities which are deemed essential to the growth and welfare of the institution and which may bring with them the necessary financing.

Also, in the "project" type of support, members of the scientific community are becoming directly and increasingly motivated toward engaging in research which is regarded as important by a sponsoring agency of the government rather than by their employer. Since most federal support is directed toward practical goals which will serve the needs of the country, there are incentives for an individual to engage in research which will receive this support and therefore may come under the heading of "mission-related" rather than "free" research.

By the way, what will happen if the ceiling on R&D funds is held more and more tightly? If we believe in substantial support for free research, with its admittedly vague and uncertain potentialities, how are we going to protect it? Will it have to depend upon income from capital funds—if so, from what sources? Or will its advocates try to oversell it by extravagant claims?

The influences that govern scientists in their choice of research and their choice of employment are more complex than ever before. Today's ivory tower is more apt to be built of reinforced concrete or stainless steel. These influences are many, some major and some detailed. For example, in addition to the competition between applied and basic research, there are considerations such as needy areas, attractive sources of funds, national or humanitarian causes, "big" versus "little" science, and deference to the plans of one's department or institution. A different kind of influence on research is represented by the following: too much assistance to thesis-writing graduate students, with an eye toward grant or contract renewal; hasty writing and issuance of research reports scanty in detail and acknowledgement; a tendency to keep a weather eye on funds for extra salary or other perquisites. Further complications are provided by administrative requirements which seem essential to management in large organizations as a means of accounting for public funds, but which distract and hamper the researcher.

But I do not wish to sound too pessimistic. As a matter of fact, I have had rather extensive contact over the past years with scientists in senior academic administrative posts and can assure you that, by and large, they understand these problems and try to hold them within manageable limits. The real danger lies in the fact that in such an extensive enterprise there are bound to be abuses. If these are not dealt with forthrightly they may spontaneously proliferate until there is clamor for formal corrective regulation.

If one were to classify the sources of influence, the first and obvious category would be money—money for projects, buildings, research equipment, salaries, and many minor perquisites. A second category would be the employing institution, in its desire for income, growth, and prestige. One would also have to list the increasing effect of personal advancement or gain associated with the positions of high responsibility, salary, and prestige which are now available to scientists.

Even science itself is providing dilemmas for an individual scientist. Should he join an interdisciplinary team in which his specialty is needed, join a large research center such as a high-energy particle accelerator installation, take part in an extensive planned program, such as oceanography or the study of pollution? Or should he remain aloof as an individual investigator? And what about his responsibility toward teaching?

Of course the consequence of all this may be the broadening out of a scientific career into one more closely integrated with society in general. This is natural enough, and surely after careful consideration most would agree that this result is desirable. My question today directly concerns the necessity for maintaining the strength and integrity of science in the face of varied opportunities, responsibilities, and distractions: How should this strength and integrity be safeguarded? If the involvement of scientists in social affairs brings with it questionable or dangerous consequences to society, then society will take steps to formulate regulations for their prevention, with possible grave effect upon science. Similarly, in science itself, if the course of science and the behavior of scientists appear to scientists themselves to be damaging to its strength and progress, then a normal reaction on their part would be the formulation of rules and regulations to prevent such abuses.

However, in order to maintain and protect the independence and creative quality of basic research in science, one should, I believe, conclude that such modes of regulation should only be attempted as a last resort, and even then as sparingly as possible. It should be clear that the most effective means of maintaining the objectives and initiative that have always characterized science is still the cultivation and retention of a strong sense of competition, cooperation, and integrity on the part of all scientists. All we need do is to continue and strengthen our time-honored traditions. But this is not going to be easy. We shall have to distinguish clearly between our conduct in our science and our behavior in the presence of issues that go beyond science alone. Judgment and objectivity are still required on such issues; the main differences are that these decisions, in contrast to science, require the weighing of opinions and pressures, as well as facts, and the attempt to make value judgments between items that are not comparable. Moreover, in the world of science, compromise has no place; in the world of affairs it must often be reckoned with, and occasionally sought.

I cannot close without mentioning a great opportunity before us which may and should become a most effective avenue for the healthy growth and influence of science. I refer to the progress made in international science programs. As is well known, science has always transcended national boundaries, and scientists of all nations have communicated and collaborated in all its disciplines. There are two categories of research for which international collaboration is especially well suited. The one includes matters of urgent public concern, and is typified by the World Health Organization and the World Meteorological Organization. Of the nature of applied research and development, these matters are, appropriately, planned and sponsored by formal agreement among governments under UNESCO. Such problems as population control, insurance against war, famine, drought and pestilence, and the development of natural resources belong in this category. In all these, science can provide a unique input, the effectiveness of which will depend directly upon the recognition of this fact by governments and people everywhere, and upon intelligent and widespread support by them.

The other category is research concerned with fields of basic research, such as geophysics and astronomy, which require concerted global observation and collection of data. Frequently this is an interesting combination of "mission-related" and "free" research. The International Council of Scientific Unions is performing meritorious service in providing a focus for these endeavors. The outstanding example, of course, is the International Geophysical Year (IGY) and its offspring—the Indian Ocean Expedition, the International Year of the Quiet Sun, the Earth Mantle Project, and the International Biological Year. Unique among these is the Antarctic Research Program, where the IGY program is continued under a 12-nation treaty, expressly and solely for purposes of scientific research.

It is in such areas that scientists are eminently qualified to plan and to operate, and it is in the highest interests of both science and government that they do so. Plans thus formulated may be submitted to their respective governments for support and any formal arrangements needed.

But beyond this, we stand at the threshold of scientific findings that will pave the way for developments of a different order of magnitude and novelty than the world has ever known. A few are already in sight—notably the exploration of space; others are as yet beyond the horizon. Some will present severe social problems; some may be dangerous; some will be extremely expensive. All will present questions for society that go far beyond the natural sciences alone; they will strongly involve the social sciences and the humanities. They will provide inspiration for the arts. To solve these problems will require many of the skills of our civilization, the utmost in statesmanship, and a general understanding and appreciation on the part of all.

The significance of these developing enterprises in science and technology, their hazards, and their excessive cost in money and manpower point to the overwhelming desirability of international cooperation. Herein lies our great opportunity as scientists—to take the lead in collaboration with our colleagues in other lands and to support our governments in furthering such collaboration.

It would be a tragedy indeed if these undertakings were to become the subject of national or sectional ambitions under conditions of unfriendly competition. On the other hand, if we can help achieve an atmosphere of collaboration, in friendly competition, we may look forward to continued healthy progress in our ideals and in our accomplishments for the future of mankind.

In the post-Sputnik era (1958 to 1965), statesmen of science were concerned with managing and planning properly the vast material success that science enjoyed. After 1965, problems of a very difficult kind became visible, and the addresses of the AAAS's presidents reflect these new concerns. After the Vietnam War shifted into high gear in 1965, various critics of society and of science-technology became vocal and increasingly evident in the media. With the rising concern for ecological problems, technology (and, in its train, science) drew considerable fire from those who saw no way out of the "ecological crisis" other than through dismantling modern industrial society. Other critics, more selective in their targets, paid special attention to what they saw as heedless, wrong-headed technical proliferation, and mounted attacks against the plans for developing the supersonic transport and lobbied in favor of automobile emission control.

Still other critics of the direction of science came from within. Radical caucuses grew like mushrooms at national meetings, and there were vocal criticisms of Defense Department support for basic research and of what was seen as the misuse of science and technology. Racism, imperialism, and militarism became subjects of heated concern, if not debate.

It was not long before scientific research began to receive bad news from its patrons. Cutbacks in government spending ended the seemingly endless prosperous years. At the end of 1970, Philip Handler, president of the AAAS, told his colleagues that during the five previous fiscal years the purchasing power of scientific support had been reduced by 20 to 25 percent. All those young people who had been exhorted and cajoled into entering scientific professions in the post-Sputnik years were having difficulty in finding jobs. "The employment prospects of the Ph.D. class of 1971 seem rather forbidding," Handler disclosed. President Nixon was reminded of what candidate Nixon had said on October 5, 1968: "[W]e must bring about a new dawn of scientific freedom and progress." But in January 1973, alarm concerning the future of federal support was widespread as the President's Science Advisory Committee and the post of Presidential Science Advisor were eliminated.

Even the success of the Apollo moon landing in 1969 did little to stem this new mood of pessimism among scientists. There had always been critics of science and technology; in fact, the arguments against science and technology had changed little in the last 50 or 75 years. Taken together with America's own time of troubles, however, the criticisms seemed, for the first time since the Depression, to wound.

Don K. Price's presidential address in 1968 lashed out against what he saw as "political reaction" on the one hand and antiscience "rebellion" on the other. The political reaction was designed to reduce basic research in favor of more emphasis upon applied science and to break down the research grant autonomy that the leaders of the scientific community had so jealously guarded. The rebellion, more serious, derives its ideology, according to Price, from Herbert Marcuse and Jacques Ellul and its dangerous power from its ability to convince intellectuals, both within the ranks of science and without, that science may be an inhumane discipline. There is one thing, however, that the reaction and the rebellion have done: they have ended the era of easy optimism—the belief in automatic progress. Science, as such, does not control policy decisions, but it bears the responsibility of helping to clarify public values, define options, and assist responsible leaders in their attempt to control the forces let loose on "this troubled planet."

The next year, in Boston, Walter Orr Roberts likewise spoke of our "imperiled" planet, but the tone of the address was very different. Just a few months before, the first Apollo manned flight to the moon had been a huge success, and that part of the scientific-technical community closest to the project was both elated about their achievement and apprehensive about the future. Roberts looked back at the pre-Sputnik period and saw "cutbacks in federal support of scientific research not unlike those occurring now." He hoped that another great thrust forward, like the one after the Sputnik "awakening," would occur. Aware of the grave problems besetting man, Roberts advocated a Global Atmospheric Research Program: "By aiming the thrust of our post-Apollo space efforts toward beneficial terrestrial applications, we can serve the tangible interests of people in all corners of the globe." He concluded that we "must turn our newly discovered skills toward the construction of world systems that make the planet earth even better than it is now for its burgeoning numbers of people. We must invent new technologies. We must commit the resources of space science . . . to the achievement of an optimum balance of man and nature."

Bentley Glass's 1970 address turned in upon the possibilities of scientific knowledge itself. Are there finite limits to scientific understanding, he asked, or are there endless horizons? His analysis pointed to the former: "We are like the explorers of a great continent who have penetrated to its margins in most points of the compass and have mapped the major mountain chains and rivers. There are still innumerable details to fill in, but the endless horizons no longer exist." Pointing to an exponential acceleration in the accumulation of scientific knowledge, Glass suspected that the rate of advance of the scientific *frontier* is actually declining. Moreover, it is obvious that the increase in numbers of scientists and technologists (doubling in slightly more than ten years) cannot be long maintained and may reach an upper level of 25 percent of the labor force. Progress, which he takes to be coupled to scientific advance, cannot continue indefinitely.

> We have almost a generation or two [Glass claimed] before
> progress must cease, whether because the world's population
> becomes insufferably dense or because we exhaust the possible
> sources of physical energy or deplete some irreplaceable resource,
> or because, most likely of all, we pollute our environment
> to toxic, irremediable limits.

But the exhaustion of progress, Glass continued, is neither inevitable nor necessarily permanent. One solution is political. We face difficult times indeed unless man forges new social systems—beyond capitalism or socialism as it now operates—that are more responsive to, and careful of the needs of humanity. A second answer is genetic. Man as he is today may not adjust. The nature and character of man will change if what we have known as progress ceases. Genetic engineering, in the interest of eugenics, may lie ahead. Perhaps mankind, Glass concluded, can escape its follies and survive the era of "no progress," in order that history may resume. Have we yet cast our die?

Purists and Politicians

Don K. Price
Dallas, 1968

Don Krasher Price, Jr. (1910–   ) *is a political scientist and public administrator whose lifelong research has been primarily in the relations between science and government in the United States. While attending Vanderbilt University (B.A., 1931), Price was a reporter and state editor for the* Nashville Evening Tennessean. *A Rhodes Scholarship took him to Oxford University, where he received the B.A. (1934) and B. Litt. (1935) for his thesis comparing British and American civil service administrators.*

After returning to the United States in 1935, Price had his first experience in government service as a research assistant with the Federal Home Owners Loan Corporation, which was in the process of developing a national housing policy. After two additional years' experience with the public administration committee of the Social Sciences Research Council, Price joined the staff of the Public Administration Clearing House, a government service research center, in 1939. Here he worked as writer, editorial assistant, assistant director, and, from 1946 to 1953, associate director. He served with the Coast Guard during wartime; after the war he worked briefly with the Bureau of the Budget, helping to formulate the legislation for the formation of the Atomic Energy Commission and the National Science Foundation. Price joined the Ford Foundation in 1953 as associate director and served as a vice-president for the next five years. In 1958, he accepted an appointment as professor of government and dean of Harvard's Graduate School of Public Administration, a position he still holds. Among other innovations, Price developed a seminar in science and public policy that has served as a model for other institutions.

Among his books are Government and Science: Their Dynamic Relation in American Democracy *(1954) and* The Scientific Estate *(1965), which explore the relation of science to the political and economic systems in the United States. Price's election to the presidency of the AAAS in 1967 was something of a rarity, since the association has had few social scientist leaders in its 125-year history.*

Sometimes the tone of a headline tells you more than the news. Last summer a New York *Times* headline read "Pure Physicists Stay That Way: Vote to Remain Out of Politics." The story was a straightforward account of the decision by the members of the American Physical Society that it would adopt no resolutions on political issues. The flavor of the headline suggested a great deal more: that the typical newspaper readers and perhaps even a good many scientists are still inclined to think that the moral obligation of a scientist is to remain aloof from policy issues and political controversy.

Since I applauded the tactical decision of the physicists but deplored the implications of the headline, it occurred to me that this apparent contradiction was worth some further thought. Perhaps it is the crux of the apparent dilemma which the entire scientific community shares with the physicists. The dilemma is ages old—the dilemma between truth and power, or, rather, between starving in the pursuit of truth and compromising truth to gain material support. But it takes its new form in the dilemma posed for the scientific community as it now comes under attack simultaneously from two sides—from a political reaction and from a new kind of rebellion.

This attack from the two extremes makes it hard for the scientific community to continue its traditional political strategy, especially since—as sometimes happens in politics—the two extremes may in effect be allies, even though superficially in conflict.

The traditional political strategy of scientists has been to keep their sights set firmly on the advancement of basic knowledge in the conviction that their mode of thinking is in the vanguard of political and economic progress, and at the same time to persuade politicians and philanthropists to support science for its indirect payoffs in power and wealth.

This strategy was based on a belief in automatic progress that had its origins in the same way of thinking that produced economic laissez-faire. Scientific knowledge, like economic initiative, could be relied on to produce progress if government could be persuaded not to interfere, except with the necessary subsidies.

But now, under attack on two fronts, scientists find this strategy harder to sustain.

On one side, the attack comes from a political reaction, which has three main purposes. Politicians want to cut down on the appropriations for research, to have more of the money spent on practical technology and less on academic theory, and to break down the degree of autonomy which the leaders of the scientific community gained a generation ago in the procedures by which research grants are distributed. On each of these points the reaction conforms to the best American tradition of the political pork barrel.

On the other side, the rebellion is a cosmopolitan, almost worldwide, movement. One is tempted to identify it with its violent and fantastic and adolescent fringe—flower power and student insurrections. Obviously, the young are the ones who charge the cops in Chicago and barricade the buildings at Columbia or Berkeley. They have to be: my contemporaries no longer have the muscle and the wind for such exertions. Today's youth are indeed the student activists, just as today's youth are the infantry in Vietnam. But it would be as much a mistake to give the student leaders credit for the ideology of the rebellion as to give the G.I.'s credit for the war plans of the Joint Chiefs.

The ideology of the rebellion is confused; you can find in it little clarity or consistency of purpose. Its mood and temper reflect the ideas of many middle-aged intellectuals who are anything but violent revolutionaries. From the point of view of scientists, the most important theme in the rebellion is its hatred of what it sees as an impersonal technological society that dominates the individual and reduces his sense of freedom. In this complex system, science and technology, far from being considered beneficent instruments of progress, are identified as the intellectual processes that are at the roots of the blind forces of oppression.

For example, André Malraux, denying that the problem is one of conflict between the generations, says that "the most basic problem of our civilization" is that it is "a civilization of machines," and that "we, for the first time, have a knowledge of matter and a knowledge of the universe which . . . suppresses man."

Jacques Ellul, one of the heroes of alienated young Europeans, presents a more systematic indictment of scientists as "sorcerers who are totally blind to the meaning of the human adventure," whose system of thought is bringing about "a dictatorship of test tubes rather than hobnailed boots."

The theme was echoed by Erich Fromm in his support of Senator McCarthy's presidential candidacy, in a public protest against the type of society in which "technical progress becomes the source of all values" and we see as a consequence "the complete alienation and dehumanization of man."

Herbert Marcuse, who is of course the favorite philosopher of the rebels, reduces the issue to its fundamental point: "the mathematical character of modern science determines the range and direction of its creativity, and leaves the nonquantifiable qualities of *humanities* outside the domain of exact science . . . [which then] feels the need for redemption by coming to terms with the 'humanities.' "

In one sense, the challenge does indeed come from the humanities. The student rebels and their faculty sympathizers, at home and abroad, are found more conspicuously in the departments of humanities and in schools of theology than in the natural sciences or engineering. If the danger comes from the humanities, however, it comes not because they are politically powerful but, rather, because, as Mr. Marcuse suggests, they may have convinced scientists themselves that science is an inhumane discipline. The case for laissez-faire vanished when businessmen themselves became aware that unregulated initiative brought depressions and economic disaster. The potential effects of the power created by modern science and technology are so obviously dangerous to the modern world—whether in terms of the cataclysm of war or the slower but equally disastrous degradation of the environment—that it would not be surprising if even scientists should wonder whether we have been reduced to these dangers by the reductionism of their system of thought.

Most scientists try to avoid thinking about this basic problem very much because they are apt just now to worry more about the reaction than the rebellion. For the reaction touches sensitive budgetary nerves in anyone who is a laboratory director or a department chairman or even an aspirant to a fellowship.

I think this choice is a mistake. The reaction is a tolerable discomfort, the rebellion a fundamental challenge—and a challenge that poses problems scientists should think about critically rather than dismiss with contempt.

It is easy and misleading to blame the reaction on the Vietnam war and therefore to sympathize with the anti-military sentiments of the rebellion. But this view overlooks the facts that two earlier wars produced more money and autonomy, not less, for science, and that the civilian agencies of government (including those with some of the most generous and humane purposes) have been more likely than the military to insist that research funds be spent on practical problems, and that they be distributed more evenly among universities and regions.

Indeed, it seems to me that the reaction mainly uses the war as an excuse, and it is hard to see how the reaction could have been so long delayed. In slowing down the rise in appropriations, congressmen were reacting naturally to the projection of curves on the budgetary graphs that lumped basic science together with engineering development. In emphasizing application, they responded to the salesmanship of scientists who told them in congressional hearings a great deal about how science would make us healthy and wealthy, and very little about how it would make us wise. And, in their avarice on behalf of their own districts and institutions, congressmen differed only in degree from scientists themselves. In these practical ways, the reaction is in the highest tradition of the English-speaking scientific world, which has always assumed that science was justified in large part by its contribution to material welfare—the tradition of Francis Bacon, who caught cold and died while trying to learn how to refrigerate poultry, and of the Royal Society, with its initial interest in "Manufactures, Mechanick practices, Engynes, and Inventions," and of Ben Franklin's American Philosophical Society, "held at Philadelphia for promoting useful knowledge."

But the rebellion is a different matter. It is the first international radical political movement for two or three centuries (I am tempted to say since Francis Bacon) that does not have material progress as its purpose. Far from proposing to use science and technology to improve the material welfare of the poor, it rejects technological progress as a political goal. Far from calling on government to distribute the fruits of technology more equitably, it denounces big organization in government and business indiscriminately. For three centuries science has worked on the comfortable assumption that it could pursue fundamental truth and at the same time contribute to human welfare and humane values. Since Bacon, revolutionary leaders have accepted the assumption and considered science to be in the vanguard of political progress. But now the rebels say that science, in its intrinsic nature, has reduced itself to an inhumane mode of thought, and our polity to an engine of oppression, and so they conclude that humane feelings demand the overthrow of the whole system, if necessary by an international rebellion.

Even though many of the young rebels call themselves Marxists, the guiding spirit of the rebellion is as much in conflict with Marxism-Leninism as with Western democracy—perhaps more so because communism believes that science can provide the basis for political values, and the New Left considers the degree of scientific influence over a political system a disaster. Communism is a system of rigorous discipline and meticulous dogma; the New Left has neither. It is more like a religious heresy, renouncing a concern for power and wealth, than like a political movement, and even its emphasis on drugs and sex is reminiscent of the antinomian rebellions of the Middle Ages.

The rebels are right when they complain of the symptoms of sickness in modern society—symptoms that afflict the communist as well as the capitalist world. We have not learned how to make our technological skills serve the purposes of humanity, or how to free men from servitude to the purposes of technological bureaucracies. But we would do well to think twice before agreeing that these symptoms are caused by reductionism in modern science, or that they would be cured by violence in the name of brotherhood or love.

As the first step toward a diagnosis of our problem, we must admit that, as scientists, we have not been very clear in the past as to the basic relation of science to politics. When the rebels charge science with destroying freedom by subverting moral values or controlling policy decisions, we cannot dismiss the charge by repeating the old principle that political authorities determine policies on the basis of philosophical or moral values, and that scientific knowledge only tells us how best to carry out those policies—that is, tells us the best means to those ends. This reply will no longer do. In Marxist countries official dogma holds that science determines the basic values, and in America many scientists have been hypocritical on the issue; they use the old formula as a defense for public relations, even though they realize that science has, and must have, a profound influence on values, and are inclined to believe that science could provide the answers to policy questions if politicians were not so stupid.

It is high time that we become more critical—instead of hypocritical—in facing this fundamental issue. As we do so, we should remember that the relationship of science to politics has at least three aspects. They are knowledge, institutions, and policy.

Let us consider knowledge first. The way people think about politics is surely influenced by what they implicitly believe about what they know and how they know it—that is, about how they acquire knowledge, and why they believe it. In traditional political systems—a few still persist in the world—issues were decided on the basis of immemorial custom, religious tradition, or the divinely sanctioned will of a ruler. Before this could change to a system in which elected assemblies could consider facts—perhaps even on the basis of scientific evidence—and then deliberately enact policies, a revolution in the nature of knowledge had to take place. That long slow revolution went along with the progress of science, and the main line of progress has of course been that of *reduction*—the change from systems of thought that were concrete but complex and disorderly, and that often confused what *is* with what ought to be, to a system of more simple and general and provable concepts.

It is clear, as Mr. Marcuse points out, that this reduction of knowledge to its abstract and quantitative bases is a calculating approach to reality that makes no allowance for humane sentiments or moral judgments. It is also clear that serious practical politicians who disapprove in theory of Mr. Marcuse may agree with him in practice, and may fear that reductionism will impair our political responsibility. For example, leading candidates for political office have charged that the Supreme Court's weakness for sociology and statistics is eroding the moral fiber of the nation, and congressmen in committee hearings have expressed concern that the new mathematical techniques of systems analysis may dominate our strategic decisions.

But it is not at all clear to me that reductionism is a threat to political freedom or responsibility. In their practical political behavior, scientists are not quite so consistent or doctrinaire. To say that science feels the need for redemption seems to me (if I may use a technical literary term in addressing a scientific audience) a pathetic fallacy. Science feels nothing. Scientists have feelings, and on political issues their feelings seem to me to be just as varied and moralistic as anyone else's.

On a more theoretical level, it seems to me that reductionism has not been pushing scientists generally toward a belief that science as such can solve the issues in which the average man is most interested, or can determine the nature of the political system. Although other branches of science admit their growing reliance on mathematics and physics, they seem no more likely than they were a century ago—perhaps less likely—to assume that they can solve all their problems by reducing their disciplines to atomic or subatomic bases.

The notion that scientific advance cuts down the freedom of the human spirit, and reduces the range of choice open to mankind, is an obsolete idea; on the contrary, every new grand simplification opens up a new range of complex questions for exploration. Man found it hard to change from the astronomical conception of a closed world to one of an infinite universe; the notion that scientific advance on reductionist principles will cut down our freedom, in either intellectual or political terms, seems to me the result of hanging onto an obsolete and narrowly mechanistic 19th-century conception of science.

Before we decide that the remedy for our present disorders is to put moral sentiments back into science, it may help us to remember that science is not the only mode of thought which has gone through a reductionist trend and then found that the simpler abstract concepts provided less specific guides to action than one might hope. If reduction is the change from complex and disorderly ideas that confuse what *is* with what men would like, to more simple and general (if not always provable) beliefs, the change in theology from polytheism to monotheism was reductionist, and so was the change from the Ten Commandments and the intricacies of the Talmud to the simpler commandment to love God and your neighbor as yourself. And in theology, as in science, reductionism brought a shocking denial that natural laws were in harmony with human righteousness: "for He makes His sun rise on the evil and on the good, and sends rain on the just and on the unjust."

If science can learn any lesson from theology on this point, it is that reductionism does not cause the political problem, nor can it solve it. For the simple law of love was taken, over the centuries, as the antinomians' justification for the abandonment of all moral laws as well as for the rigorous moralism of Calvinist Geneva and the Spanish Inquisition, for the anarchy of the hermits and the Ranters as well as for the ruthless tyranny of the Byzantine emperors.

The trouble with reductionism, as far as politics is concerned, is not that it gives *all* the answers to the important issues but that it gives hardly any. I suspect that the current attacks on science come less from those who have always feared it than from those who were frustrated when they tried to put too much faith in it. To them, it was another god that failed. Science is quite impartial in debunking idols—its proudest claim is that it is always debunking itself.

If we are concerned with political freedom we cannot concern ourselves only with the theory of knowledge. Reductionism in science is not the real problem. We do harm not by reducing science to its mathematical bases but only by reducing men to a concern for nothing but science. As we ponder the political status of science it may help us to recall that freedom of religion resulted less directly from the reformation of theological thought than from the competition of dissenting churches and from changes in the political system itself. That brings us to the second aspect of the relation of science to politics—institutions. And so we must face the question of whether a scientific and technological establishment, or the aggregate of scientific and technological institutions, is a threat to freedom, especially because of its intimate alliance with a bureaucracy managed on scientific principles.

At the same time that science has been reducing knowledge to fewer and simpler general concepts, society has been expanding the number and the variety of the institutions that develop and apply that knowledge. From the traditional community ruled by a priest-king, combining in one set of institutions political power and the preservation and transmission of traditional knowledge, has been evolved the complex structure of modern society. This process of *specialization* has separated from the center of political power various more or less autonomous institutions that are then permitted to operate according to their own functional requirements.

The fundamental basis for the freedom of specialized institutions is that the public recognizes that they can do their particular job better for society if they are not immediately controlled by those who hold ultimate political power. The business corporation can be more efficient, the scientific laboratory can be innovative, if it is granted substantial autonomy. And the same principle works, within limits, within the formal structure of government itself; it is the justification not only for a nonpolitical judiciary but for a professional diplomatic or military service and for a civil service run on merit principles.

But the free institutions' role in serving society is not merely to be more efficient within their specific functions. It is also to serve as a source of independent criticism of those who hold power. It is, in short, to prevent centralization of authority. Scientific and technological competence is so necessary today for understanding the complex programs of government that scientists who are employed by institutions outside the immediate executive hierarchy have an important role to play in criticizing official policy and checking centralized power.

Obviously, an institution can be more surely free of political influence if it deals with pure science and shuns the competition for power. But absolute purity is a delusion. It is a delusion partly because every institution needs material support and cannot isolate itself from the society that supports it. Even more important, absolute purity is a delusion because it is a refusal to serve one of the essential purposes of an independent and nonpolitical institution, that of providing some independent standards of criticism of public policy.

You can resolve the dilemma in one of two ways. If your approach is doctrinaire, you can try to resolve it by forcing the competing elements together within a single institutional system. Politics and religion are obviously related, so church and state cannot be separated. Economic and political power are related, so the state must own the means of production. Political decisions must be made scientifically, so science must provide a theory of politics and a methodology for deciding public issues, and then must be controlled by the state. That way, of course, lies totalitarianism.

But if you are sensitive to the danger that any single doctrine or theory may be perverted in the interests of power, you will take a more pluralistic and more discriminating approach. You can distinguish between different types and degrees of political involvement on the part of nonpolitical institutions; even more important, you can distinguish between what it is prudent and effective for an institution to do and what that institution's members are free to do in their capacity as private citizens or as participants in other institutions. (I hope it was this line of reasoning, more than any fundamental distaste for politics, that led the American Physical Society to abstain from political resolutions.) A member of a church may also be a member of a political party and need not expect both institutions to play the same roles. A professor in the university may also be a consultant to a research corporation or a government agency and a member of a scientific society. His freedom to play different roles in these different institutions—and to defend the autonomy of each institution against the others—is one of the most important safeguards of freedom in modern society.

Independent institutions are not, of course, the fundamental basis of freedom. Their independence comes from their roots in the way people think and what people believe. You will not want to let a university or scientific society function free from governmental direction if you think its work will immediately determine the major political decisions of the day.

We believe in free academic and scientific institutions not because we consider them irrelevant to practical political concerns but because we tacitly understand that their type of knowledge does not directly and clearly provide the answer to any complex political issue. Does this contradict the power of the reductionist approach that has given science its great effectiveness in dealing with practical as well as theoretical problems? I think not. Reduction in knowledge and specialization in the definition of institutions and their roles go hand in hand. Just as the zoologist or botanist may admit the great contributions that biochemistry and biophysics have made to biology and still see that tremendous problems remain at the more complex levels of organization, to be dealt with by different modes of thought, so the politician (and his scientific adviser) may make full use of analytical science and yet be left with difficult problems of synthetic judgment in making his decision.

The type of thought that, in the style of the Marxist dialectic, rejects traditional dogma in favor of reductionist science, and then tries to make science the basis of a new dogma, is not reductionist, but only dogmatic. Reductionist knowledge provides no rationale, and no rationalization, for centralized authority. Like the specialized institutions in which it is developed, it tends to be a check on general political power, an impediment to sovereignty rather than a tool of tyranny. Reductionism and specialization have indeed biased our political system toward some of the practical abuses of power that the rebels deplore, but they have done so not by creating a centralized system. On the contrary, they have so greatly strengthened the productivity and power of specialized concentrations of economic wealth and technological competence that our general constitutional system is incapable of controlling them.

This brings me to the third aspect of the relation of science to politics—policy, or the definition of public purpose by responsible authority.

As a complex civilization has developed its system of knowledge by *reduction*, and its institutions by *specialization*, its policy has moved over the centuries toward *generalization*. The purposes of politics have broadened from the tribe to the feudal community to the nation, and are beginning dimly to be perceived in terms of world interests; they have broken down the rigid lines of caste and class, and are beginning to transcend differences of race. With almost as much difficulty, the general purposes of responsible politics must now try to control the specialized functions and institutions of government in the general interest.

This movement toward political concern for all men, and toward the sharing of power with them, was perhaps made possible by the other two aspects of politics—the reduction of knowledge to a more effective scientific basis and the transfer of specialized social function away from the general system of sovereignty to institutions less concerned with power and more with material welfare. Without new techniques of communication to let men share ideas from place to place, and new techniques of production to give them enough material goods to share, the broadening of political concern would have been impossible.

I am also inclined to believe that this broadening of public purpose was encouraged by that earlier form of reductionism, the theological reductionism that slowly and partially converted religion from a complex of local superstitions to a broader and simpler faith. As far as the general evolution of public policy is concerned, the processes of reduction toward simpler and more fundamental ideas in science and in religion have had similar effects.

But I must qualify this assertion of faith with a cynical concession. Science has an intellectual advantage over religion: a reductionist science comes out with grand generalities in the form of mathematical equations that the layman reveres because he cannot understand them; a reductionist religion comes out with grand generalities in the form of platitudes that only embarrass the layman because he thinks he understands them all too well. For example, the tough-minded ghost writers for one of our leading politicians, I am told, were always annoyed at being required to put into each of his speeches a reference to what they called BOMFOG—their derisive acronym for the Brotherhood of Man and the Fatherhood of God.

As scientists we are apt to take pride in this distinction: even pious people, unless they are simpleminded, can laugh at BOMFOG, but nobody makes fun of $E = mc^2$. But such pride is ill-founded. If we ridicule BOMFOG, it is not because we do not believe in God or human brotherhood; indeed, the more we believe, the more we are likely to see that such belief does not solve practical political problems, and that a politician who appeals to such abstractions for self-serving purposes is absurd. It seems obvious to us that $E = mc^2$, while it may be the fundamental equation of atomic energy, does not tell us even how to make atomic bombs, much less how to get international agreement against their use; no politician would win votes by using a basic scientific formula as an incantation.

But this is a parochial idea. We may not make a political slogan out of a scientific concept, but others do. We find it hard to imagine the political quarrels that took place in Russia over the scientific philosophy of Mach or Einstein, or to understand how Soviet scientists give credit for their discoveries to Marxist-Leninist doctrine, and Chinese scientists give credit to the thoughts of Mao. But, at least in Russia, the more sophisticated scientists react to the scientific dialectic the way we react to BOMFOG—with an appropriate mixture of reverence and ridicule.

If, as Americans, we have escaped the communist habit of muddling scientific theory with political practice, we cannot claim too much credit.

We had been inoculated, so to speak, by the English-speaking historical culture against the translation of the great simple truths into practical policy. We had tried that under Puritanism—under Oliver Cromwell and John Winthrop—and had had enough. So Jefferson, as clearly as Burke, was against the Worship of Reason in the French Revolution, and T. H. Huxley opposed Comte's conversion of science into a political dogma—the dogma (which Lenin later enforced) that diversity of opinion was no more to be tolerated in politics than in chemistry.

With respect to knowledge and institutions, politics becomes more civilized as it moves in the analytical direction—toward reduction and specialization. But policy is a synthetic process: generalization requires more than analytical skills. Indeed, it demands special care with respect to analysis and specialization, not to prevent but to control and use them, and not to be misled by thinking that any one type of basic knowledge or institutional skill will solve the problems of a complex political organization. Reduction is the prescription for basic knowledge, but reductionism—taken neat—can be poisonous for policy.

America is not entirely free of the idea that some scientific formula will guarantee our political salvation. The president of the AAAS gets frequent letters outlining such schemes. If I were not too honest to steal such secrets from their authors, I could tell you how to provide unlimited energy without cost, and thus eliminate poverty, and how to remove all feelings of hatred and aggression, and thus guarantee universal peace. But it is typical, I think, that most of those American scientists whom their colleagues consider crackpots are interested, not in basic theory or ideology, but in gadgetry—in finding gimmicks to cure the world's ills.

This taste for the so-called "practical," of course, the crackpot shares with his fellow countrymen. In America, we are not dialectical materialists, only practical materialists. We do not convert our science into political faith—only our technology into business profits. We do not make our political theory into a revolutionary crusade; we only assume that technical assistance and more calories will make peasants contented, and that B-52's are cost-effective in pacifying jungle villages, and that welfare payments will remove racial hate in our urban ghettos.

The philosophers who blame such blunders on scientific reductionism—who believe that the mathematical and fundamental approach to knowledge is the basic flaw in modern politics—are themselves reducing the problem to a more abstract level than is useful. We get into political difficulties less because our method of knowing is wrong than because we put too much confidence in specialized programs and institutions and show too little concern for the processes of government that relate those specialties to general policy. It is true, of course, that many political controversies are over meaningless issues or insoluble problems, and new "technological fixes" (as Alvin Weinberg calls them) are often useful ways out. But this approach will work best if it is tried by some responsible authority who is thinking about the problem as a whole, as a part of the general political system; it can be disastrous if it is peddled to politicians by a special interest in the business or bureaucratic world that is concerned only with increasing its own profits or professional influence.

To deal with any public issues of any consequence, we need to bring science and politics together in all their aspects. We need more precise knowledge. We need more effective institutions. And we need both the will and the competence required for the synthesis of general policy. Of these three, the most difficult is the policy aspect, for generalization cannot be reduced to precise techniques, or delegated to a specialized profession or institution.

But synthesis and analysis are not incompatible processes of thought, any more than facts and values are totally separated from each other. The new techniques for the analysis of complex systems developed by mathematicians, physicists, economists, and other scientists have become the most powerful tools for the critical study of the components of policy, and hence for the development of general policies.

You cannot synthesize a sensible policy unless you have first analyzed the problem. Reductionism is not the enemy of humane political thought; it is the first practical step toward it. To take both steps is hard work, and requires the scientist to share the complexities and uncertainties that harass the politician, and to join in compromises that offend the purist in either science or morals.

From these uncertainties, the human mind is tempted to seek refuge in phony reductionism—the new rebels reducing the complexities of politics to the simplicities of moral feeling, the scientists taking shelter in the purity of research. Both these paths to purity are like BOMFOG—you feel obliged to respect them, but the trouble comes in putting them into effect.

What is wrong with the purists, on both the moral and scientific sides, is not that their objectives are evil but that they tackle the problem at the wrong level of abstraction. In the United States we are in no danger of using science to deny political freedom, or of rejecting BOMFOG in favor of a theology that would support a caste system. But there is a real danger, it seems to me, that the two types of purists—the scientist and the moralist—will withdraw from public affairs and leave responsible political authority without support against the powerful combination of technological skill and special industrial and bureaucratic interests.

For example, take the Institute for Defense Analyses. IDA is a prime target for the new rebels; to them it symbolizes the corruption of the purity of scholarly institutions by military power. IDA is also not very popular among theoretical scientists; it represents the kind of applied work with government support that does little for pure science. Yet IDA was not created in the interest of irresponsible military power. On the contrary, it was a part of the effort to give responsible civilian political authority the ability to control the competing special military interests. The constitutional authority had always been there, but without the special knowledge or the special institutional controls needed to make that power real, and hence to make possible the synthesis of the independent missions of our Army, Navy, and Air Force into a general policy.

Even before 1961, IDA was one of the tools the Secretary of Defense used as an aid in the synthesis of general policy. There was no antithesis, here, in either theory or practice, between, on the one hand, reductionist knowledge and specialized staff institutions and, on the other, an effort to make general policy supreme over special technological interests.

In opposition, we saw officers from the most powerful and independent segment of American bureaucracy, the career military services, supported by industrial clients who disapproved on principle of any not-for-profit corporation, rise to denounce the whiz kids in the research corporations and the Office of the Secretary of Defense. The use of mathematical and scientific techniques to deal with military policies such as strategic plans and weapons systems, was a cold and calculating and heartless approach, they said, to what ought to be an affair of the heart—a vocation to be followed on moral rather than quantitative principles. Or as Admiral Rickover put it, "The Greeks at Thermopylae and at Salamis would not have stood up to the Persians had they had cost effectiveness people to guide them."

I find much of Admiral Rickover's critique of our overemphasis on technology and bureaucracy refreshing—especially coming from an admiral. What other admiral would ask, "Does man exist for the economy or does the economy exist for man?" and charge that the "larger bureaucratically administered organizations" in which most Americans now work, as a result of the industrial and scientific revolutions, "are in every respect the obverse of a free society"?

But I doubt that this rhetoric, which ought to endear the admiral to the new rebels, really advances our understanding of the nature of freedom in modern society. Whenever a powerful special interest begins to appeal to basic moral or philosophical principles in an effort to escape subordination to general policy, we are entitled to be skeptical, if not cynical. The new purists in morals and in science who join with rebellious segments of the Air Force and Navy in attacks on IDA and the Office of the Secretary of Defense are in much the same position as the contemporary religious fundamentalist who becomes an ally of reactionary industrialists by seeing social security, the income tax, and the regulation of business as the work of Godless communism.

In the current state of the world the question of whether scientific societies should pass political resolutions is a trivial tactical issue; the community of science needs to look to its broader strategy.

In this strategy the idea of scientific purity—of avoiding involvement in political compromise—was once a useful notion. It helped to free science from the teleology of the earlier philosophers, and scientific institutions from the obligation to work on practical problems as practical men defined them.

This reductionist strategy, while protecting the freedom of scientific institutions, did not slow down the practical application of science in political systems that had shaken off feudal or bureaucratic constraints in an era of optimism about material progress.

But the new rebels are right in thinking that that era of optimism—that blind faith in automatic progress—has ended.

That optimism misled Western thought in two ways for a century or two after the Enlightenment.

After the French Revolution there spread eastward through Europe and Asia the optimistic notion, stemming from the Enlightenment, that science, by perfecting our philosophy and our values, will teach us how to revolutionize society and eliminate the corruptions of politics; in its Marxist form, that notion proposed to let the State itself wither away.

After the American Revolution, the pragmatic West came to a less doctrinaire but almost equally optimistic conclusion: that the advancement of science would lead to the progress of technology and industry and an increase in material prosperity, and to a withering away of governmental interference with private initiative.

The rebels are right in being pessimistic about such notions. I do not think they are even pessimistic enough. To me it seems possible that the new amount of technological power let loose in an overcrowded world may overload any system we might devise for its control; the possibility of a complete and apocalyptic end of civilization cannot be dismissed as a morbid fantasy.

And the rebels are far too romantically optimistic in their remedy. Mere rebellion to destroy the existing order—mere purposeless violence to upset the establishment—assumes that those who gain power by violence will be nobler and more generous in purpose than those who now try to hold together the delicate web of civilized institutions.

If scientists wish to maintain the freedom of their science and, at the same time, play a rational and effective role in politics, they need to adopt a strategy that is more modest in its hopes for the perfectibility of mankind and more pessimistically alert to the dangers of power—not only power that is obviously political but the power that calls itself private as well. They should start by acknowledging in theory what in the United States we have always taken for granted as a practical matter: that reductionism in scientific knowledge, while it may provide the fundamental advances in scientific theory, does not alone provide the answers in the realm of policy, or the basis for a political ideology.

If this point is clear, no one will need to take seriously the charge that the scientific mode of thought is a fundamental threat to humane values. The threat comes not from the theoretical reductionism of science but from the very pragmatic reductionism which assumes that applications of advanced technology are automatically beneficial, or that we are always justified in granting special concentrations of technological and industrial power freedom from central political authority.

If everyone understands that science, as such, does not control policy decisions, scientists will then be free—and, in my view, will be morally obliged—to devote their synthetic as well as their analytic skills to the formulations and criticism of policies by which the nation may control technology and apply science in the public interest.

In an era which is beginning to be alert to the threats posed by modern technology to the human environment, the role of science in politics is no longer merely to destroy the irrational and superstitious beliefs which were once the foundation of oppressive authority. It is, rather, to help clarify our public values, define our policy options, and assist responsible political leaders in the guidance and control of the powerful forces which have been let loose on this troubled planet.

After the Moon, the Earth!

Walter Orr Roberts
Boston, 1969

Walter Orr Roberts (1915–), *astronomer and geophysicist, was educated at Amherst College and Harvard University, where he took his Ph.D. in 1943. Roberts spent the next 15 years at the High Altitude Observatory at Climax Station, Colorado, as supervisor, director, and, later, founder of its research center. Aided by the observatory's coronagraph, Roberts identified and described solar spicules and other aspects of the solar corona, later becoming interested in the influences of solar activity on terrestrial phenomena. By 1955, this interest drew him to meteorology and climatology; since then, Roberts has been an active proponent of interdisciplinary work in the atmospheric sciences. He has been professor of astrogeophysics at the University of Colorado since 1957 and served as director of the National Center for Atmospheric Research, Boulder, Colorado, from 1960 to 1968. He was president of the AAAS in 1968.*

Shortly after the turn of the 17th century, Galileo Galilei gazed with wonder upon lunar craters, which were revealed for the first time by a new invention, the telescope. It was, he said, a "most beautiful and pleasurable sight." More important, however, was the fact that Galileo's observations opened a whole new era of observational astronomy. The science, philosophy, and poetry of the subsequent centuries show the imprint of man's audacity in discovery about the heavens.

Now, 360 years after Galileo's first overwhelming view of the rugged terrain of the moon, four men have walked on those very mountains of the moon. Today you and I can look for ourselves at the strange, gray, porous-looking rocks the astronauts brought back from the moon. It was an exciting moment for me, a few weeks ago, when I visited the public museum at NASA, Houston, to look from bifocal proximity at these drab rocks and to ponder about the scintillating reflections that came, apparently, from irregular glassy lumps in small holes at many points of the rock surface.

Less than a decade ago, on 25 May 1961, President John F. Kennedy made the moon landing a national goal. The United States, he said, "should commit itself to achieving the goal, before this decade is out, of landing a man on the moon and returning him safely to the earth." I must confess that I was afraid that the President had asked us to accomplish the impossible. The task demanded technological attainments of unprecedented complexity and difficulty. It required supreme skills in a field of engineering where we had much yet to learn.

But the goal has been attained, not once, but twice. Its accomplishment is not only a triumph of science and technology but also one of the truly great adventures of mankind.

To visit other worlds has been one of the long-unfilled dreams of mankind. Only a few years after Galileo's first drawings of lunar terrain, the astronomer Johannes Kepler, discoverer of the laws that govern planetary motions, wrote a fascinating and poetic science-fiction, *Somnium*, in which earthlings were transported to the moon over the bridge of shadows that briefly spans the quarter-million miles to the moon during a lunar eclipse. There they met strange, tough-hided creatures. Many other writers followed Kepler's lead, creating fanciful tales of space adventure.

Cyrano de Bergerac wrote in 1656, with irrepressible but unscientific exuberance, of a marvelous jet-propelled spaceship in which he traveled to the moon and the sun. His contraption, scarcely less imaginative than Apollo 12, looked like a combination of a globe, an oversize filing cabinet, and a huge sail; it was propelled by jets of steam, and rose with the morning dew. On the moon, the prophet Elijah showed Cyrano how to use magnetic propulsion.

Then, as now, there were critics. Samuel Johnson was speaking of Cyrano, among others, when he wrote caustically, in *The Adventurer* for 10 April 1753, of ideas like space travel:

> A voyage to the Moon, however romantic and absurd the
> scheme may now appear since the properties of air have been
> better understood, seemed highly probable to many of the
> aspiring wits in the last century, who began to dote upon their
> glossy plumes and fluttered with impatience for the hour
> of their departure.

But the skeptics did not deter either the enthusiasm of the fiction writers or the efforts of the serious proponents of flight into free space. Jules Verne's prophetic *From the Earth to the Moon* had a great impact on world expectations for space travel.

The accelerating drive toward mastery of space brought to the scene men like Tsiolkovsky, Goddard, Oberth, von Karman, and von Braun. An instrument of war, the German V-2 rocket, made a giant stride, on 3 October 1942, by lifting off at 14 tons and flying its payload 50 miles into space. The threshold of the space age was first crossed, however, by the electrifying Sputnik I, a Soviet satellite of 184 pounds payload fired into earth-circling orbit on 4 October 1957.

I shall never forget that Friday evening. I had been present in Barcelona, the year before, at a planning meeting for the International Geophysical Year, and I had heard the chairman of the Soviet IGY Committee announce his country's preparations for launching a scientific satellite. But I was not prepared for the world-stunning impact of the event, when it occurred. I hurriedly rearranged some powerful radio receivers at our laboratory that evening, and then listened and tape-recorded the sounds of that first 20-megahertz space transmitter as it passed over the United States in the early evening of that first day.

When the Soviet Union launched Sputnik II, just a month later, with the dog Laika as passenger, the world was aware that we had entered a new age. Men in every country suddenly sensed the pace of modern scientific and technological advance. The fact that this marvelous feat of engineering had been achieved by the Soviet Union and not by the United States was noted everywhere, and it was a great blow to American self-confidence. The impact on world opinion and on domestic opinion was remarkable. Considering the magnitude of the event, however, I do not think the reaction surprising. Our self-esteem was not helped, in early December, when our country's first effort to orbit a man-made satellite, Vanguard I, ended in a gigantic explosion and fire.

The events had been preceded in our country by cutbacks in federal support of scientific research not unlike those occurring now. But the reaction to the setback in prestige went far deeper than simple restoration of the budget cuts. The Sputniks stirred an American self-assessment that affected everything from science to the priority ranking of national goals and domestic political attitudes. Even the very assumptions underlying our educational system came under grave questioning.

The effect of the Soviet achievements on our world image and our self-image persisted, even though Vanguard II and Vanguard III, launched but a few months later, achieved their aims. The U.S. Army's Explorer satellites of early 1958 were even more prominent and scientifically successful. I well remember, for example, the IGY Special Committee meeting in Moscow in the summer of 1958, where Homer Newell presented exciting new results about the trapped radiation belts, relayed by cable from James A. Van Allen, who had discovered the belts and mapped out their main features from Explorer-series satellites.

The months that followed the first Sputniks saw the creation of NASA, and the origins of America's great push into space science under distinguished civilian leadership, and in full view of the world. The achievement of President Kennedy's goal, last July, with the first moonwalk, was an incredible event. It was not only the realization of a national objective but the fulfillment of man's agelong yearning to visit another world. It is estimated that the July moon venture was seen or heard, live or in replay, by one-fourth of all the people of the earth! I watched with a small group before a flickery TV in a room in the Sejong Hotel in Seoul, Korea. Outside, the normally bustling traffic of the street was nearly halted, as if during curfew. Later in the day everyone—university professors, policemen, elevator girls, waiters, taxi drivers, shopkeepers, and even pedestrians on the street—hailed the event and congratulated the Americans.

It is interesting, from today's perspective, to compare actual events with the expectations of only a few years before Sputnik I. For example, the noted Russian scientist V. V. Dobronravov, who then headed the Interplanetary Navigation Committee of the Soviet Air Force Aero Club, in July 1954 wrote a timetable of expected space accomplishments. It went like this: 1965, unmanned earth-orbiting satellite, low orbit; 1975, three-man, earth-orbiting space ship; 1980, manned trip to moon orbit, no landing; 2000, manned rock-collecting field trip to the moon. Dobronravov was not a poorly informed prognosticator; the pace and priority of the Soviet and American efforts have simply exceeded earlier expectations. Moreover, the gigantic group efforts applied to the problem by both nations have demonstrated the almost irresistible power of large group efforts to achieve earnestly sought goals that are technologically feasible. Dobronravov's timetable was more a political-economic prediction than a technological-feasibility forecast, and it was grossly conservative.

NASA's policy of conducting all of its flight operations in full view of the world brought admiration at home and abroad. By so doing, we demonstrated that we were not afraid to show our failures as well as our successes. The world was watching at our launches, for better or worse. The record of safety and success is fantastic, in spite of the tragedy of "the fire" which cost the lives of three of our astronauts, including Edward White, who talked to the AAAS of his orbital space walk at our Berkeley meeting in 1965. The United States can now look with pride upon its space program. Its success has made a firm imprint on the thinking of people in every part of the earth.

Many men and animals, Soviet and American, have orbited in space to bring knowledge of the physiological and psychological effects of sustained weightlessness and other aspects of space flight. Unmanned probes have brought new details of the planets Venus and Mars. Soon we shall have the answer to the ancient riddle of whether there is earthlike life on Mars, or indeed any recognizable life at all.

Waste products of Soviet and American launches—hundreds of pieces of man-made space debris—wander in earth-circling or deep space orbits, or lie inert on the lunar surface. Astronauts Conrad and Bean revisited our unmanned Surveyor lunar laboratory in their Apollo 12 pinpoint landing and brought back fragments of the craft for analysis of the deterioration wrought by more than 2½ years in space. The moon's Sea of Tranquillity and Ocean of Storms bear crisscrossed footprints of earthmen that will be eradicated, in all likelihood, only after millions of years of cosmic particle bombardment. Overboots, film cassettes, used packages, tools, backpacks, cameras, and other discarded items dot the lunar landing sites. Among them are deployed seismographs, corner reflectors, and other still-operating experiments sending data to the earth. And there are, too, plaques and mementos honoring Soviet and American spacemen who died seeking to learn the mysteries of the cosmos. In this exotic place these artifacts of man will probably outlast the civilization of earth itself.

Space technology has, alas, also vastly enlarged the arsenals of the United States and the U.S.S.R., bringing to reality hitherto fanciful modes of military surveillance, communications, and weaponry. No point on the earth is more than minutes from the possibility of atomic attack by orbital or ballistic weapons carriers. The major powers stockpile ever more powerful and more deceptive ballistic missiles, antiballistic missiles, and anti-antiballistic missiles. The end of the spiral is not in sight. It is quite certain that, even if a civilian space program had not emerged, weapons would be poised for ballistic or space orbit. The V-2 demonstrated, if anyone needed such proofs, that man's ingenuity can be turned as well to instruments of war as to tools of peace.

In the years between Sputnik I and Apollo 12 many scientific or practically useful advances have occurred as a direct result of space research, and many nations besides the Soviet Union and our country have participated in these developments. Unmanned probes have explored the solar wind, have monitored the sun in hitherto inaccessible wavelength ranges, have registered the interplanetary magnetic field deep in space, have discovered the magnetosphere of the earth and its trailing tail, and have uncovered totally unexpected properties of the trapped radiation of the Van Allen belts. The science of cosmic-terrestrial relations has been profoundly altered by these discoveries. Earth-synchronous and low-orbit satellites have given us a new and highly useful view of earth weather, and have entered into routine global service for weather forecasting. Indirect probing of atmospheric temperatures has given astonishingly successful results, promising major advances in global weather measurement. Communications satellites are being used with increasing reliability and economy for a rapidly growing range of applications. Satellites are also finding significant uses in astronomy, geology, hydrology, forestry, oceanography, navigation, geodesy, and aeronomy.

Each new flight to space brings fascinating new mysteries to be unraveled. Apollo 11 revealed the glassy lumps in the small craters, perhaps evidence of a prehistoric superflare of the sun, and gave us samples of the strange, colorful, tiny glass spheres abundant in the lunar dust. The seismograph left by Apollo 12 revealed, when the spent lunar escape module was crashed back against the moon's surface, a weird, long-lasting, one-cycle-per-second oscillation that first rose slowly to a crescendo, then declined, remaining detectable for 50 minutes. These and the many more mysteries of this strange, dead companion-world of the earth will stimulate new experiments and new discoveries, some perhaps with direct significance for our world.

There can be no doubt that President Kennedy's goal has brought us great returns, both tangible and of the spirit. Before we have completed the seven additional Apollo lunar explorations that are scheduled through 1973, we will have placed still more sophisticated scientific experiments on the moon, and we will have obtained field data from a wide range of additional locations on the moon.

With this great goal nearing realization, it is necessary, now, to look to the next steps in space. What are the alternatives before us? Shall we seek to send a man to Mars, as some advocate? If we do, should we set a timetable for reaching the goal? Shall we abandon our hard-earned skills in space and turn our scientific-technological drives toward other kinds of goals entirely? Hard choices face us. We have grave problems in the international realm, of which the Vietnam war is but one. Can any nation be secure in a world that is part rich and part poor? Can the earth's population spiral ever higher without ominous consequences for all? At home, pollution of the environment, decay in the core city, chaos in the air traffic lanes, disenchantment of the young with our values and our choices—these and other problems crowd upon us, and make demands on our resources. Where does space research stand among men's needs in these areas?

I am not able to answer these hard questions directly. However, in the remainder of this address I want to express my personal views about the best next steps in space for our nation and the world. I am aware, of course, that thoughtful men have expressed quite different conclusions on these matters, and that I cannot hope for immediate or universal acceptance of my position. But I do hope that there will be a sustained and thoughtful national debate before our decisions are irrevocably sealed.

I believe that the time has come, at this moment of great achievement for the United States, for us to take a bold new step in space. I propose that this nation call upon the Soviet Union to join hands in space, with a jointly conducted, earth-oriented space program that will put the new-found Soviet and American skills in space to work for the direct benefit of man, and with a maximum of international cooperation. There are many effective ways in which this can be done, with vast potential returns both in the short run and in the longer term.

By aiming the main thrust of our post-Apollo space efforts toward beneficial terrestrial applications, we can serve the tangible interests of people in all corners of the globe, and we can also greatly advance international cooperation and international understanding. There are extraordinary opportunities, in a properly conceived space program, for multinational planning and execution of specific experiments and investigations oriented toward the peaceful uses of outer space, both for pure science and for direct earth-oriented applications. By working jointly with the Soviet Union we can make the effort a matter of joint prestige—a matter of man against the unknown, rather than of Americans versus Soviets.

There is, moreover, a magic in the perspective from space that makes our planet appear as the hospitable, good earth that it really is for man. Nearly all of those who have been in space have a new view of earth. From lunar orbit, the earth is home—not Houston, or the United States, or North America, but "the good earth," as the crew of Apollo 8 expressed it in greeting us from the first lunar orbit on Christmas day just a year ago. What better step, at this time, than to internationalize our efforts in space, and to direct them largely toward improving the abode of man and achieving peaceful relationships among nations.

Yuri Gagarin, the Soviet cosmonaut who lost his life in that nation's space effort, spoke articulately to this point in May 1966:

> It is very important, in my opinion, to enlarge international
> cooperation for mastery and use of cosmic space, so that each
> flight of man into the cosmos, each launching of stations
> and of scientific laboratories in space will serve all mankind
> in the name of life and of peace.

If man's yearning for the secrets of other worlds is great, his yearning for a peaceful and plentiful world is even greater. With imagination, we could just perhaps make the space program of the future a major contributor to the realization of these added dreams of all mankind.

Let me turn to some specific suggestions. In my view, there would be great benefit to our nation and to the world if we were immediately to announce our intent to initiate, as our major post-Apollo thrust in space, a series of joint Soviet-American space programs also open to other nations as collaborators. These efforts could begin almost at once, with the space research areas where there are needs for satellite launches by both countries and where there is some degree of parity between the state of advancement of the present operations of the two countries. Let me give three specific examples of areas in which I believe we could organize joint programs in the decades to come.

1) *Applications satellites.* Remote sensing in visible and infrared wavelengths affords a number of areas where there would be virtue in the creation of coordinated Soviet-American experimental programs, with assignment of space on vehicles to experimenters of both countries. This would permit intercalibration and intercomparison of the different sensors and techniques on identical information fields.

There would, of course, be some problems. I am sure that both countries would face difficulties over the "interface" between classified and unclassified work; these could certainly be solved. Questions concerning compatibility of readout systems, telemetry resolution, and the like would also arise. It seems to me highly probable that it would not take long to achieve the goals in spite of the problems.

The first areas for joint work would probably be those where substantial cooperation has already been the rule, such as in remote sensing of the atmosphere's temperatures for meteorological purposes. Certain areas of solar physics and of interplanetary particle and magnetic field probing hold similar promise of mutual benefits from joint work, and have a similar history of close international cooperation.

I do not propose, here, to describe the many important areas of earth-oriented satellite applications that offer promise of advances in pure research and in benefits of practical importance to man. All of them are areas suitable for a joint space program. Some have greater promise and practicality than others. A detailed study has recently been concluded and published by the U.S. National Academy of Sciences, with attention given to the probable practical uses to which earth-oriented satellites can be directed in coming years, and with recommendations for next steps. The fields embraced include such areas as meteorology, forestry and agriculture, geology, oceanography, communications, navigation and traffic control, economic analysis, geodesy, and cartography. The sights were perhaps not set high enough in this study, but it would be a good beginning from which to approach the Soviets.

The interest in such applications is no less intense in the Soviet Union than it is here. Participation in discussion of earth-oriented satellite applications at the International Astronautical Federation, for example, has been enthusiastic, and detailed. Strong Soviet enthusiasm is found in areas relating to agriculture and climatology. I have also heard Soviet scientists discuss with great optimism the possibilities for manned orbital laboratories staffed with teams skilled in earth-resource research, from agriculture to mineralogy to air conservation, and I discuss the prospects for a cooperative space laboratory below.

2) *Joint U.S.-Soviet Venus and Mars exploration programs.* In planetary exploration programs, Soviet-American cooperation might be especially fruitful. I maintain that the thrust in Mars exploration should not be "a man on Mars," even if no date is set for the landing. In my view the thrust should be a solely scientific exploration, in which the best and the most economical means of achieving the desired scientific aims are chosen. This would, in all probability, rule out serious effort, at least now, to carry out a 2-year manned Mars landing, just as NASA has tentatively worked out mission details. The fearful jump in difficulty that a manned Mars landing entails, as compared to a manned moon landing, would bring manned planetary flight costs to levels that I hope appear prohibitive to others, as they do to me.

On the other hand, a joint Soviet and American program of scientific exploration, while not directly earth-oriented, seems fully justified at this juncture in history—among other reasons, as a means of satisfying man's curiosity about the universe. But it must not receive the lion's share of the space manpower and dollars, or rubles.

There is also great merit, in planetary science programs, in bringing in other international participants. In many instances mutual benefits could be attained simply by comparing plans for, and coordinating launches of, Soviet and American rockets destined for Venus or Mars, with each nation choosing its own collaborators. The present work of the international body COSPAR (Committee on Space Research) leans this way already, but does not go far enough. If our space program were specifically vectored toward cooperative space research, and if the Soviet Union were similarly oriented, the work of COSPAR would "graduate" to a far higher level of significance.

The accidental coincidence of Soviet and American Venus probes a year or so ago permitted a considerably better understanding of the temperature structure of that planet's atmosphere than would otherwise have resulted. Such benefits should occur by design, and not by accident. The Soviets' strength in the development of advanced, unmanned planetary probes will probably make this area of cooperation attractive to them, because they will be in a position of essential equality from the outset.

3) *A joint manned space laboratory.* One of the high-priority future items for space is, in my view, a manned space station, with earth-oriented research instrumentation and personnel. I can visualize a not-distant future time when such space stations will form an integral part of world monitoring networks, with teams of skilled observers studying hurricanes, tropical ocean-atmosphere energy transformations, ocean current flow patterns for fish-migration analysis, air pollution drift, and spread of insect pests and plant diseases, assessing water reserves in the world's watersheds, and making a host of other terrestrial studies.

The world would benefit from priority efforts toward development of such space stations, served by space shuttles and space tugs and manned for considerable periods by trained crews. I believe that such stations should be developed under international auspices, and should be manned cooperatively by individuals from a variety of nations, or by international public servants. The United States could make a great contribution to world understanding by announcing, at this time, a program to internationalize our efforts to build, man, and operate research and operations stations in space. It would be possible to put international teams into the stations that will follow directly after the prototype U.S. space station now scheduled for launch in 1972. It can be done in Antarctica; why not in space?

There are, of course, problems that come about from the military uses of space that neither the U.S.S.R. nor the United States will want to compromise by internationalizing all space knowledge. But the separation that already exists between NASA and the military space program should make this a minimal obstacle on our side. I believe that there is no better time than now to try. Moreover, such an effort might trigger better progress in arms limitation and in open-skies policies for the world. From a spaceship it is hard to see the logic of political boundaries. A new concept of sovereignty of nations might turn out to be the most important ultimate product of joint space stations.

Until more nations of the world can participate actively in the benefits from space research, the true potential of our new-found space skills will be at least partially wasted. I believe, therefore, that it is appropriate, as a part of our plans for a post-Apollo space program, to seriously consider creating, in developing regions of the world, new kinds of international space research centers oriented toward the exploitation of space science and technology for the practical benefit of local, but broad, geographical regions. A center for Africa, for example, would make good sense. Similarly, one for Latin America.

The focus of such centers would be on the research and development that would lead to improvements in exploration for mineral resources, control of agricultural pests, improvement of crops, effective use of forest resources, development of marine resources, prediction of weather and climate—all through the use of space tools and techniques, imaginatively coupled to earth-based research and development.

The fields of promise for such centers embrace, for example, geology, weather, forestry, agriculture, fisheries, computer science, environmental pollution control, and communications (including communications in the areas of education, medicine, entertainment, and commerce).

A center dedicated to this purpose, for example, might resemble, in some important ways, an international version of the laboratory with which I am affiliated, the National Center for Atmospheric Research, at Boulder, Colorado. It would have a strong component of pure science, but there would also be skilled engineers and technicians on its staff, and it would carry out applied research and development as well. Its research facilities should be of high quality and power. And it should fulfill, as well, an important educational and training function, though it should not be a university.

In such a center there should be various groups and divisions dedicated to problem-solving related to specifically chosen social and economic needs of the continent or region in question. There should also be pure research activity in the fields most relevant to the specific problems. For example, there would need to be, for exploitation of space-sensing of oil resources, geologists familiar with the principles of petroleum geology and also knowledgeable about the geology of the regions involved. Obviously, there would be need for a good deal of training in the space-based techniques, and one might also expect that the center would become a focus of new discoveries regarding ways to instrument and use satellites for such purposes. In particular, the center would be a direct readout point for the remote-sensor data from space satellites programmed to provide the fundamental inputs for regional research.

Each center would need to be under the direction of a highly skilled scientist or engineer who was dedicated to its goals and who was in close communication with the satellite-development and launching agencies abroad. The staffing of such centers would probably have to involve a mix of senior scientists from highly developed countries and from the local region embraced by the center.

A large fraction of the staff of such a center would obviously have to be visitors, but there would have to be very outstanding resource people on hand permanently. Especially is this so in regard to the computer facility that would be needed at such a center.

The creation of appropriate management and financing might be a ticklish business, as would be some of the necessary agreements regarding on-the-ground backup for the space-based sensing. Another difficult area would be the matter of "user taxes" or other questions that would arise when major resource development in a particular country derived, at least in part, from the work of a center. There would be jockeying about the location of the center. Nonetheless, these problems can be solved, I am sure.

I see no reason why the United States and the Soviet Union could not, cooperatively, provide systematic help to such centers. This help could consist of the design and launch of satellites for the resource programs, as well as the provision of compatible readout capabilities. Most important of all, however, would be scientific and technical consultation. If the prosperity and effective operation of such international centers for the peaceful uses of space were a significant aspect of our space program, we could provide the solid kind of support that would greatly enhance the probability of their success.

The most interesting and important earth-oriented space program of the future, save perhaps for the communications program, is, to my mind, one in the domain of atmospheric research and applications. I should say, in passing, that the atmosphere-ocean envelope of earth and certain aspects of the solid earth must be regarded as a single geophysical entity. Thus, physical oceanography and atmospheric dynamics are both essential components of atmospheric research, and the sensing tools deployed for the atmospheric research programs must measure the oceans as well as the atmosphere.

In spite of the amount of domestic emphasis that has been given the Global Atmospheric Research Program, it has not achieved its merited degree of public or official visibility. It should, in my view, be elevated to the status of a major objective of the U.S. space program for the decade ahead. Again, the world cooperative aspects should be emphasized. Internationally, planning for the Global Atmospheric Research Program is moving fast, and there is remarkable interest and enthusiasm. But the United States should do far more to sustain the momentum and priority of this critically important program.

The Global Atmospheric Research Program merits priority attention for the following reasons:

1) Its underlying aim is the benefit of all mankind through improved knowledge of weather and climate.

2) It is made-to-order for international cooperation, and in fact cannot come about without extensive cooperation that will involve many of the underdeveloped countries in the use of simple tools and techniques to give regional data for a global network.

3) It excites the interest of the general public as do few scientific or technological questions—especially since it includes work on many important aspects of weather and climate modification, both deliberate and inadvertent.

4) Space technologies are essential to its success, though it also makes strong demands upon conventional meteorology.

5) Detailed studies have been carried far enough so that we know fairly well its probable dimensions, costs, and products. There are enough gambles to make it exciting.

6) It has its roots in earlier U.N. resolutions introduced by the United States for the peaceful uses of outer space.

7) The Congress, by resolution, has indicated its interest in it as a national goal.

For these reasons, I urge that it be made a vital element of this nation's post-Apollo efforts in space.

The greatest thrill I can imagine for myself is to stand on the moon's surface and to look back from the harshness of the lunar landscape to the luminous, hospitable earth. From that vantage point, I believe, I could view the earth in its oneness. There I could better understand that indeed all mankind properly shares in the pride of attaining a lunar landing. After all, it is science and engineering, the common heritage of all mankind, that made it possible! All nations and races have contributed to the unbroken threads of knowledge that comprise science.

From space, the earth must indeed appear a rare and beautiful place in the vastness of the universe. I sometimes gain this sense of the emptiness and scale of the cosmos when I fly my plane over the lightless reaches of western Nevada on a dark, clear night and see the Galaxy stretched out overhead.

The past steps in space have required extraordinary bravery and emotional stability on the part of the men of our space program, who understood, as the public only vaguely does, the hazards of the task. To take the next great step in space will require no less skill and knowledge, equally steadfast courage, and a parallel sense for innovation. The next step in space must be directed toward the earth. We must turn our newly discovered skills toward the construction of world systems that make the planet earth even better than it is now for its burgeoning numbers of people. We must invent new world technologies. We must commit the resources of space science, directly and indirectly, to the achievement of an optimum balance of man and nature on this magnificent but imperiled planet.

Science: Endless Horizons or Golden Age?

H. Bentley Glass
Chicago, 1970

Hiram Bentley Glass (1906–), *a geneticist and administrator at the State University of New York at Stony Brook, was born in 1906 at Laichowfu, Shantung, China, to American Baptist missionary parents. He grew up in China and received his early education there, coming to the United States in 1923 for his higher education. After receiving the B.A. degree at Baylor University, Glass taught high school biology and physics in Texas for two years before resuming biological studies at Baylor (M.A., 1929) and at the University of Texas, Austin (Ph.D., 1932). With a National Research Council fellowship in genetics, Glass did postdoctoral research at the University of Oslo, the Kaiser Wilhelm Institute in Berlin, and the University of Missouri. Glass was an instructor of biology at Stephens College, Columbia, Missouri (1934 to 1938) and at Goucher College, Baltimore (1938 to 1947). In 1948, Glass joined the faculty at The Johns Hopkins University, where he carried out extensive studies on genetic drift in human populations. Other research areas Glass tackled were Rh factors, suppressor genes, and the history of genetics. In 1965 he became vice-president of academic affairs and Distinguished Professor of Biology at SUNY, positions he still holds.*

A prodigious writer whose bibliography extends to over 200 journal articles and books, including Science and Ethical Values *(1965) and* The Timely and the Timeless *(1970), Glass has served during his career as editor of* Quarterly Review of Biology, Science, Biological Abstracts, *and the* McCollum-Pratt Institute Symposia *(nine volumes). Glass' long-standing interest in science education led him to push for the revamping of high school biology studies; after six years as chairman of the Biological Sciences Curriculum Study (BSCS), Glass and his collaborators produced three biology courses now used by over half of the high schools in the United States. He was president of the AAAS in 1969.*

At the end of each year, and even more customarily at the close of every decade, we are wont to look back and summarize our gains and our losses, and to look forward in the hope of forecasting as well as possible what the next year, the next decade, or even the next century may hold for us. For the scientist it seems particularly appropriate, since the future of science itself—its rate of advance in various directions and the roadblocks or morasses ahead—will likely be mirrored in the rates of technological change and consequently in the attainment of human well-being or the advent of catastrophe. One need not be a prophet to attempt an analysis and critique of scientists' views about the the future of science, and these may themselves sharpen our vision of the problematical years that lie ahead.

Scientists themselves often seem little concerned about the matter. They are concerned about the quantitative measurement of all sorts of phenomena except that of the growth, stasis, or decline of their own collective activity. Young scientists, in particular, seem to find it sufficient to harbor a personal liking for scientific exploration and discovery, without much worry about the magnitude of the edifice they are constructing, the depth of the mine they are excavating, or the extent of the continent they are invading. With advancing years one commences to wonder about the relative position of one's own work in a broader, more extended perspective. Thus one is led to try to evaluate present scientific knowledge not only in comparison with the past but also the future of man's understanding of nature. Then the question must be faced: Are there finite limits to scientific understanding, or are there endless horizons?

I have chosen two phrases for my title to represent the extremes in the spectrum of belief in the future of science—the one, the view of limitlessly expanding knowledge and of infinite bounds; the other the view that scientific knowledge, like our universe, must be finite, and that the most significant laws of nature will soon have been discovered. The Golden Age is thus the age of perfected knowledge, of consummated applications, and of social stasis. The first of these phrases is the title of a book written in 1946 by Vannevar Bush about the future of science. The phrase "Golden Age" is also from the title of a book, *The Coming of the Golden Age*, written by Gunther Stent and published in 1969. They thus epitomize the view of two eminent scientists, one a physical scientist and engineer and the other a geneticist and molecular biologist. Each of them was writing at the close of a period of unparalleled development in his respective area of science. In 1946, Vannevar Bush could contemplate from the vantage point of his recent work as director of the wartime Office of Scientific Research and Development what Karl T. Compton called "the greatest mobilization of scientific power in the history of the world." Not only could Vannevar Bush think with just pride of the advanced weaponry which had been created but, even more important, of the harnessing of scientific talent in new ways through operations research and systems analysis. His own inventions, the network analyzer and the differential analyzer, were harbingers of the Age of Computers. He foresaw the problems involved in converting atomic energy from a military secret jealously guarded by a few nations to a worldwide basis of power production open to all nations but subject to the hazards of power politics. He recognized the difficulty in promoting scientific advance in the atmosphere of secrecy, in the face of demands for "classified" information. He was foremost among those who planned and advocated national support of basic scientific research, which he felt had lagged during World War II, and he was one of the fathers of the National Science Foundation. It may thus have been quite natural for him to see ahead of us the "endless horizons" of scientific knowledge, the greatest challenge to moderns for whom the West has been reached, the world's continents settled and the oceans crisscrossed, for whom in fact the excitement of the geographical frontier has vanished.

Like Bush, Gunther Stent was writing at the end of the greatest advances ever made in his own scientific field. Genetics, as a science, did not even exist before the opening of the 20th century. Molecular genetics, which Stent identifies as the third and fourth phases of the science, its so-called "Dogmatic" and "Academic" periods, began only in 1953 or thereabouts, and yet in the ensuing years it has added a greater knowledge of the chemical basis of heredity and its mechanisms than in all the years before, and seems, according to Stent, to be well into a final period of diminishing returns. More and more scientists, publishing more and more papers, fill in missing details and extrapolate knowledge in quite predictable directions. The great conceptions, the fundamental mechanisms, and the basic laws are now known. For all time to come, these have been discovered, here and now, in our own lifetime. They can never be discovered again unless man loses his scientific heritage; and even so, they cannot be discovered again for the first time. They can only be reexhumed, refined, or modified. Like Newton's laws of motion, like the Copernican solar system, like the origin of species, or like relativity and quantum mechanics, they are now the known, not the unknown—and by just so much is the extent of unknown Nature diminished.

Like Roderick Seidenberg, with whose book *Posthistoric Man* Gunther Stent is apparently unacquainted, Stent expects that the cessation of scientific advances will ultimately lead to an end of technological and social progress. Both Seidenberg and Stent acknowledge their debt to Henry Adams in charting the dynamics of progress and predicting its ultimate cessation. Where Seidenberg pictures the final state of society as posthistoric, in the sense that, having achieved ultimate knowledge and its applications to man's life in numerous ways, society will crystallize into a changeless pattern, Stent, foreseeing the virtual end of an exponential increase in human power in a very few centuries, and the consequent end of the arts and sciences, predicts a Golden Age, when man will live in a way "not very different from a re-creation of Polynesia on a global scale." It is worth quoting his description of this final state of society:

> The will to power will not have vanished entirely, but the
> distribution of its intensity among individuals will have been
> drastically altered. At one end of this distribution will be a
> minority of the people whose work will keep intact the
> technology that sustains the multitude at a high standard
> of living. In the middle of the distribution will be found a type,
> largely unemployed, for whom the distinction between the real
> and the illusory will still be meaningful and whose prototype is
> the beatnik. He will retain an interest in the world and seek
> satisfaction from sensual pleasures. At the other end of the

spectrum will be a type largely unemployable, for whom the boundary of the real and the imagined will have been largely dissolved, at least to the extent compatible with his physical survival. His prototype is the hippie. His interest in the world will be rather small, and he will derive his satisfaction mainly from drugs or, once this has become technologically practicable, from direct electrical inputs into his nervous system. This special distribution, it will be noted, bears some considerable resemblance to the Alphas, Betas, and Gammas in Aldous Huxley's *Brave New World*. However, unlike Huxley, I do not envisage this distribution to be the result of any purposive or planned breeding program, but merely a natural population heterogeneity engendered mainly by differences in childhood history. Furthermore, in contrast to the low-grade producer roles assigned to Betas and Gammas, beatniks and hippies will play no socioeconomic role other than being consumers.

As far as culture is concerned, the Golden Age will be a period of general stasis, not unlike that envisaged by Meyer for the arts. Progress will have greatly decelerated, even though activities formally analogous to the arts and sciences will continue. It is obvious that Faustian Man of the Iron Age [we of today imbued with the idea of progress] would view with considerable distaste this prospect of his affluent successors, devoting their abundance of leisure time to sensual pleasures, or what is even more repugnant to him, deriving private synthetic happiness from hallucinatory drugs. But Faustian Man had better face up to the fact that it is precisely *this* Golden Age which is the natural fruit of all his frantic efforts, and that it does no good now to wish it otherwise. Millennia of doing arts and sciences will finally transform the tragi-comedy of life into a happening.

Here, then, are diametrically opposed views of the future of man which grow out of the contemplation by the scientist of the achievements of science itself. What are the basic assumptions on which these views are founded? It seems to me that the implicit assumption of greatest importance relates to the finiteness, or alternatively the infinitude, of possible knowledge. No one really questions, I believe, that even now we may be like little boys on the shores of a vast ocean, tossing pebbles into the waves. What remains to be learned may indeed dwarf imagination. Nevertheless, the universe itself is closed and finite, or at least bounded to our knowledge by the radius of the light-years since its beginning, about 10 billion years ago; and our telescopes have now plumbed space almost to those limits of the observable. Their growing knowledge of the universe has led scientists to believe more and more firmly that the laws of nature have universal applicability. Matter is composed of the same particles and elements everywhere. Radiant energy moves with the same speeds and has the same characteristics everywhere. Local differences are explicable in terms of local conditions and past history.

The uniformity of nature and the general applicability of natural laws set limits to knowledge. If there are just 100, 105, or 110 ways in which atoms may form, then when one has identified the full range of properties of these, singly and in combination, chemical knowledge will be complete. There is a finite number of species of plants and of animals—even of insects—upon the earth. We are as yet far from knowing all about the genetics, structure and physiology, or the behavior, of even a single one of these. Nevertheless, a total knowledge of all life forms is only about 2×10^6 times the potential knowledge about any of them. Moreover, the universality of the genetic code, the common character of proteins in different species, the generality of cellular structure and cellular reproduction, the basic similarity of energy metabolism in a species and of photosynthesis in green plants and bacteria, and the universal evolution of living forms through mutation and natural selection all lead inescapably to a conclusion that, although diversity may be great, the laws of life, based on similarities, are finite in number and comprehensible to us in the main even now. We are like the explorers of a great continent who have penetrated to its margins in most points of the compass and have mapped the major mountain chains and rivers. There are still innumerable details to fill in, but the endless horizons no longer exist.

It seems to me that Vannevar Bush would not really disagree with the foregoing analysis. When one examines his book, *Endless Horizons*, for the expected discussion of the challenge of scientific advancement to mankind and for some consideration of the actual meaning of the title, one seeks in vain. A great deal of attention is given to specific technological advances and to the possible further developments of the future, but the nearest to a discussion of the theme implied by the title is in two brief sections of the essay "As We May Think." These are headed, respectively, "Endless Trails" and "Horizons Unlimited." Yet the endless trails turn out to be only searches through the coded items stored in a vast mechanical memory, or "memex"; while the unlimited horizons unfold only in the form of specific predictions of various ways in which science may in the future develop the production, storage, and retrieval of information, or discover new extensions of our human senses. Whether the horizons are really endless is an idea not only unanswered, it is not even examined. On the other hand, in his final essay, "The Builders," Vannevar Bush likens science to the quarrying of material for an edifice and the process of its construction. ". . . The material is exceedingly varied . . . the whole effort is highly unorganized [and] . . . the edifice itself has a remarkable property, for its form is predestined by the laws of logic and the nature of human reasoning. It is almost as though it had once existed. . . . Parts of the edifice are being used while construction proceeds. . . . The workers sometimes proceed in erratic ways. . . . On the other hand there are those men of rare vision who can grasp well in advance just the block that is needed for rapid advance on a section of the edifice to be possible, who can tell by some subtle sense where it will be found, and who have an uncanny skill in cleaning away dross and bringing it surely into the light. These are the master workmen."

Are these not the images of one who believes in the finite extensibility of science? Is an edifice with a predestined form infinite in extent? It cannot be affirmed that Bush's metaphors and similes unquestionably imply the finiteness of nature and of scientific understanding, but is that not the most reasonable implication? If so, was the bold title, *Endless Horizons*, not itself a metaphor, never intended to be taken literally, but supposed merely to imply that from our present viewpoint so much yet remains before us to be discovered that the horizons seem virtually endless?

If one grants that the universe itself is finite and that its laws too are universal and finite in number—or even if the universe be infinite but its nature and its laws universal—then with each new phenomenon discovered and explored, with each new law confirmed, there is an approach to the finite limits of scientific knowledge. In that case, it is less important to note the absolute bounds of knowledge at the present time than to examine the rate at which, in the past century and a half, our scientific knowledge has been expanding.

By whatever parameters one may choose to select, the accumulation of knowledge has been accelerating exponentially during this period, with a doubling time of 12 to 15 years. As I have described in a fuller study of this phenomenon in a recent book, *The Timely and the Timeless*, the number of scientific periodicals, the number of scientists publishing or working, and the number of scientific papers have been doubling almost every 10 years. It is of course dangerous to equate the growth of scientific knowledge with growth in the number of scientists or their publication of papers. Yet I suppose, although I cannot demonstrate it satisfactorily, that there must be some relatively stable mathematical ratio between the total number of scientific contributions published in a given period and the number among them of highly important, significant "breakthroughs." Derek J. de Solla Price, writing in *Science Since Babylon*, suggests that the "stature" of science, which I take to be equivalent to the frontier or attainment of science, doubles at a much slower rate than the increase of scientific information in the gross. He estimates that it has doubled, in the modern period, about every 30 years, the rate being maybe slower because of the cumulative nature of science, which like a pyramid or pile of stones must increase in volume as the cube of any increase in height. Such matters have not been studied carefully, but I suspect that the advances in technology which are basic to progress are more directly related to the rate of this advance of the frontier than to the actual total increase of scientific information, as if an 8 percent per annum technological advance would correspond to a 2 percent per annum rate of advance of the scientific frontier, that is, to a doubling time of 35 years for the latter. If our technological level as well as our total scientific information are now 256 times what they were at the beginning of the 20th century, our scientific "stature" is now but four times as great as then.

It would be most interesting if our students of the sociology of science, who are all too few in number, were to demonstrate that this rate is itself declining, as I suspect. It might be taken to indicate not only that the *average* originality of scientists is declining as the number of scientists expands, but also that it is in fact becoming more and more difficult, as scientific knowledge grows, to make a totally new and unexpected discovery or to break through the dogmas of established scientific views. Be that as it may, the amount of extant scientific information, trivial as well as important, is currently increasing at a rate related to the absolute number of scientists, technologists, and laboratory workers in the population. I have pointed out that in the United States this rate, which produces a doubling in somewhat over 10 years, cannot long be maintained, since it is much faster than the rate of population growth for the country and already constitutes about 20 percent of the entire professional labor force. I therefore suggested that the upper limit for the proportion of "scientists and technologists" in the entire professional labor force may perhaps be about 25 percent. Further absolute gains would seem to depend upon the increase in size of the total population, an increase already slackening, or a possible increase of the professional component in the total labor force. Here the limiting factor may well be the requirements for a modicum of intelligence and of advanced education and training on the part of every scientist or technologist.

Both historian and scientist agree that progress, insofar as it can be defined and measured, must be defined in terms of man's increasing power. This was the theme of Carl Becker, in *Progress and Power*, who humorously summed it up in the following words:

> All that has happened to man in 506,000 years may be symbolized by this fact—at the end of the Time-Scale he can, with ease and expedition, put his ancestors in cages: he has somehow learned the trick of having conveniently at hand and at his disposal powers not provided by his biological inheritance. *From the beginning of the Time-Scale man has increasingly implemented himself with power.* Had he not done so, he would have had no history, nor even the consciousness of not having any: at the end of the Time-Scale he would still be (if not extinct) what he was at the beginning— *Pithecanthropus erectus*, the Erect-Ape-Man. Without power no progress.
> . . . The significant fact is that the human race, so far from having any aversion from power, has at all times welcomed it as a value to be cherished. Look where we will along the Time-Scale, we see men eagerly seeking power, patiently fashioning and tenaciously grasping the instruments for exerting it, conferring honor upon those who employ it most effectively.

Stent, influenced particularly by Oswald Spengler, adopts the concept of the archetype Faustian Man, whose boundless will to power locks him "in an endless strife with his world to overcome obstacles, conflict, to his mind, being the very essence of existence." Thus, while both Becker and Stent accept the dictum of J. B. Bury that the *idea* of progress is scarcely older than the 17th century and did not really become a prevalent social theory until the beginning of the 19th, they virtually identify *all* history with the occurrence of progress through increasing power.

As long as new instruments of power are multiplied and the command over new sources of power is increased, progress may continue. But scholars and scientists alike are dubious. Says Becker:

> But it is conceivable, even probable, that the possibility
> of discovering and applying new sources and implements
> of power will in the course of time gradually diminish, or even
> be altogether exhausted. In that event the outward conditions
> of life will change less and less rapidly, will in time become
> sufficiently stable perhaps to be comprehended, sufficiently
> stable therefore for a relatively complete adjustment of ideas and
> habits to the relatively unchanging body of matter-of-fact
> knowledge of man and the outer world in which he lives.
> In such a stabilized and scientifically adjusted society the idea
> of progress would no doubt become irrelevant as progress itself
> became imperceptible or non-existent.

Roderick Seidenberg pushes his vision of a stable society and a completely adjusted man to the ultimate extreme. When the organization of society has proceeded to its final crystallization "in a period devoid of change, we may truly say that man will enter upon a posthistoric age in which, perhaps, he will remain encased in an endless routine and sequence of events, not unlike that of the ants, the bees, and the termites. . . . Man may likewise find himself entombed in a perpetual round of perfectly adjusted responses. . . . Man will hasten along his predestined way under the illusion of attaining his freedom on even higher levels of existence, while actually sealing his fate by all the devices his dominant intelligence can command." The emotions, he thinks, will have atrophied. Even consciousness, becoming no longer necessary in the absence of tension or unstable equilibrium, will evaporate.

No rate of scientific development equal to that of the past half century can be long maintained. Henry Adams pointed out that the actual doubling times are not so important as the simple existence of an exponential rate of increase, since in any case the acceleration of the process will eventually lead to staggering, and indeed impossible, magnitudes. Do not we all remember the story of the Arabian merchant who contracted with the Sultan to work for one grain of wheat the first day, two the second day, four the third day, and so doubling with each day to the 64th day, corresponding to the squares on their chessboard; and how he bankrupted the kingdom? In real increases occurring for a time at exponential rates there is, as Stent emphasizes, a positive feedback. The more offspring are produced, the more adults there will be to contribute to the next generation. The fission of one bacterium into two provides the basis for the next doubling. The recognition of progress in the economic, social, and political condition of man led to the expectation, beginning about the end of the 18th century, that progress not only was possible but was to be expected, and this expectation led men to push for further progress.

"Evidently," Stent says, "the element of positive feedback embodied by progress is that the rate at which man can gain more power over the outer world is the greater the more power is already at his disposal." Yet, the argument continues, ". . . this very aspect of *positive* feedback of progress responsible for its continuous acceleration embodies in it an element of temporal self-limitation. . . . If one examines one by one the parameters conceivably relevant to estimating the rate of progress, such as world population and energy consumption, per capita income, or speed of travel, one must conclude that none of them is likely ever to exceed some definite bound." The further we have advanced on the accelerating curve, the closer we inevitably come to the time when limiting factors will curb the growth and bring it into some degree of stasis, or equilibrium, or possibly decline.

We may find reason to disagree with various details of these pessimistic visions of man's future, but can we honestly set aside the conclusion that *progress*, in the sense of ever-growing power over the environment, must soon come to an end? We may for a long time to come retain our emotional responses, which to so great an extent depend upon hormonal balance stabilized in the early phases of mammalian evolution, long before man was man. I do not accept Seidenberg's supposition that in the keen battles between reactions on the basis of intelligence and reactions on the basis of emotion, intelligence will win the upper hand. Seidenberg seems to have mistaken the basis of emotion with instinct. True, in vertebrates such as birds or even mammals, instincts appear often to operate through the machinery of the hormones, but in bees, ants, and termites instincts depend fully as much upon the inborn connections of neurons, in other words, upon the same fundamental mechanisms as intelligence. The distinction to be drawn is between the inborn patterns of the nervous system, which develop irrespective of the external situation, most often in vertebrates before birth, or hatching from the egg, and the malleable patterns which can in some way be modified through experience and learning. Stent certainly does not envisage a decline of emotionality—quite the contrary. His beatniks and hippies, who will comprise the majority of future human populations if he is right, will react to life even more wholly than now on the basis of emotionality.

That is a side issue. Let us keep clearly in mind that the primary basis of agreement of Becker, Seidenberg and Stent, and perhaps even of Bush, is that progress cannot continue indefinitely. Indeed, so awesome is already the accelerating rate of our scientific and technological advance that simple extrapolation of the exponential curve shows unmistakably that we have, at most, a generation or two before progress must cease, whether because the world's population becomes insufferably dense, or because we exhaust the possible sources of physical energy or deplete some irreplaceable resource, or because, most likely of all, we pollute our environment to toxic, irremediable limits. Many scientists have in recent decades examined these processes and have tried to flag the runaway express. The present, more general outcry, daily growing stronger, against unlimited population growth and heedless pollution of the environment offers a slight ground for hope. The prime difficulty is that so many persons, not only in the highly industrialized countries but even more among the peoples of the underdeveloped countries, now see their hopes for the future bound in with the continuation and extension to all mankind of the progress hitherto limited to a few fortunate lands and people. The momentum of these processes is furthermore such that measures to apply the brakes to population growth or to reclaim and preserve our environment, even if the measures are firm and effective, will take at least one generation—say until the year 2000—to reach full effect.

To say the least, capitalism has not found a way to survive on the basis of a simple replacement economy. The pressure to keep the gross national product growing—in real, not inflationary, terms—comes from all elements of our society, from labor as well as employer and industrialist, from natural scientist as well as economist. (True, a few rumblings of dissent from this point of view have been voiced.) Dependence upon the profit motive, which has in the past stimulated the unparalleled growth of Western Europe and the United States, just as surely presses society into overexploitation.

Socialism, which, as it now operates in other segments of the world, is but a modified governmental capitalism, seems no better able to avoid over-exploitation than does pure individualistic capitalism. No socialist country has met the population problem as successfully as highly industrialized, capitalist Japan. The depletion of the world's resources flowing from economic imperialism is as much a characteristic of socialism as of capitalism. The pollution of the environment proceeds with perhaps even fewer checks in a socialist land than in a Western democracy, where the outcry of protesters is freer and the influence upon legislators more direct.

Man indeed faces hard times unless a new social and economic system, far more responsive to human needs and far more foresighted than in the past, can be invented. What is coming to be called technology assessment is one example of what is needed: a complete systems analysis of all the long-range effects and side effects of each technological alteration, before it is unleashed. Our local, state, and national governments are hopelessly unprepared to exercise such a function. Our legislators and bureaucrats are uninformed and unaware, for the most part, of the crisis facing the world today, and they are perhaps universally incapable of conducting the kind of analysis required. What they could do, if they would, would be to establish in sufficient numbers and strength the agencies to carry out the studies required; and they could unquestionably, if they would, assess strict penalties upon all, high or low, who waste our resources and needlessly pollute our environment.

In Czechoslovakia, as I have previously informed members of the AAAS, I discovered the existence of an Institute for the Study of the Biological Landscape. It was a response to an overwhelming technological disaster that occurred when, during the industrialization of Slovakia after World War II, a great mill was constructed in a narrow valley of the Tatra Mountains, where an atmospheric inversion layer penned down the fumes, heavy in sulfur dioxide, arising from the combustion of local coal. Every living thing in the valley had been killed, and the workers, so I was told, at times had to wear gas masks and live many miles away, at a great loss of time. In the United States—and in every other country, whether heavily industrial or newly entering the industrial phase—we need a profusion of such agencies for the study not only of the environment and its alteration by technological innovation, but also of the systems for predicting and regulating change. Let me suggest, at the risk of grave misunderstanding, that in future histories of the world the decade of the 1960's may be known not significantly for the miserable Vietnam war but as the time when man, with unbridled lust for power over nature and for a so-called high standard of living, measured by the consumption of the products of an industrial civilization, set in motion the final, speedy, inexorable rush toward the end of progress.

I am not a hopeless pessimist, however. I have already suggested one way in which mankind might avoid the debacle. These agencies to curb irrational exploitation, widespread in all countries, could mutually reinforce one another. They might form the basis of a most effective United Nations network. In any case, what the United States itself most needs at this juncture is a second Vannevar Bush who could organize and direct a comparable national effort in peacetime as Bush did in time of war. It is consequently disheartening to observe that the technological developments most needed for the kind of systems analysis required, and which Bush foresaw in 1946, are not yet available. His great "memex" system, which would replace private files and public libraries, which would code all extant information on every subject for instant retrieval, and which would provide endless cross trails of reference to stimulate fresh experimentation and analysis, is still unrealized. There seems to be little or no reason why such a system could not be constructed today. The computers, now far exceeding in capacity for storage and rapid retrieval what Vannevar Bush had dreamed of, are here. What is still lacking is the compatibility of different systems and the actual realization of the conception. The machines, in fact, are far superior to the skill of the human programmer who feeds our mechanical memory banks. In consequence, the very accumulation of disorganized scientific information is burying us daily deeper in a profusion of unevaluated, unused knowledge. Scientific information has increased a hundredfold since 1900, but most of it is already forgotten. The next age of scholarship will no doubt promote processors and analysts who need only to delve in the mountains of extant scientific and technological literature for forgotten and uncomprehended items of knowledge.

There is one other way in which the crisis provoked by an uncontrolled, exponential increase can be surmounted. Every biologist who deals with the population growth of bacteria, mice, or men knows that exponential growth not only must come to a halt when one or another factor of the environment becomes limiting, but also that subsequently a variety of fates may ensue. The population may die out altogether if it cannot adjust rapidly enough. Toxicity may increase, for example, owing to the wastes of the population itself, to such an extent that even in the presence of ample food, all organisms perish. Or fish, in a polluted lake deprived of sufficient oxygen, may die by millions. On the other hand, the population may level off toward a condition of stability, in which proliferation and exploitation of the environment are balanced against the capacity of the environment, especially the food supply, to regenerate. Or, perhaps most interesting of all, after a period of such stabilization the once-limiting factor may become no longer the limiting parameter of the environment. Then the population may set out on a new cycle of exponential growth.

If man learns in time to regulate his numbers so as not irreparably to pollute his environment, if he learns to utilize ordinary rock in place of rare metals, or to synthesize out of abundant plant materials most of what he needs, it is possible that the limiting factor may become the present supplies and sources of energy. Fossil fuels are exhaustible, perhaps within a century; nuclear energy from fission depends also upon scarce and limited amounts of uranium or artificial transuranium elements. Man's future resources of energy may thus depend upon controlled nuclear fusion. That might indeed initiate a new phase of exponential growth of human *power* (not necessarily of population).

In the new, far more regulated society of man which will inevitably be forced upon us by our exponential rates of increase, the present genetic types of man may not all permit a happy adjustment. The nature and personality of man must change, too, if progress is no longer to be our chief goal and ambition. The once sacred rights of man must alter in many ways. Thus, in an overpopulated world it can no longer be affirmed that the right of the man and woman to reproduce as they see fit is inviolate. On the contrary, if my own additional child deprives someone else of the privilege of parenthood, I must voluntarily refrain, or be compelled to do so. In a world where each pair must be limited, on the average, to two offspring and no more, the right that must become paramount is not the right to procreate, but rather the right of every child to be born with a sound physical and mental constitution, based on a sound genotype. No parents will in that future time have a right to burden society with a malformed or a mentally incompetent child. Just as every child must have the right to full educational opportunity and a sound nutrition, so every child has the inalienable right to a sound heritage.

Human power is advancing with extraordinary rapidity in this realm of control over the genetic characteristics of the unborn. Perhaps, as Carl Becker so pregnantly stated, our race, far from having any aversion from power, will welcome this power too, will seek it, fashion it, and grasp it tenaciously. Unlimited access to state-regulated abortion will combine with the now perfected techniques of determining chromosome abnormalities in the developing fetus to rid us of the several percentages of all births that today represent uncontrollable defects such as mongolism (Down's syndrome) and sex deviants such as the XYY type. Genetic clinics will be constructed in which, before long, as many as 100 different recessive hereditary defects can be detected in the carriers, who may be warned against or prohibited from having offspring. Preliminary efforts to synthesize genes and to introduce a sound gene by means of a carrier virus into a child or fetus bearing only the defective allele at the same position in the chromosomes are promising. They may make it possible not merely to correct a genetic defect while leaving intact in the individual the defective gene to be passed on, but actually to substitute the sound gene for the defective gene in the reproductive cells of the treated person. This procedure will be unquestionably most effective if carried out during the early embryonic stages of development, or in the just-fertilized egg itself before it has begun its cleavage into numerous cells. Hence we must look with expectant attention at the startling progress that is being made in the laboratory of R. G. Edwards at Cambridge University, England, and in a few other places, since the number of biologists engaged in such studies is still astonishingly low.

Dr. Edwards and his collaborators have succeeded in obtaining considerable numbers of oocytes from human females, culturing them in media that permit their maturation, fertilizing them with fresh spermatozoa, and observing normal development of the embryos so produced to the blastocyst (hollow ball) stage at which they normally become implanted in the wall of the mother's uterus. The way is thus clear to performing what I have called "prenatal adoption," for not only might the selected embryos be implanted in the uterus of the woman who supplied the oocytes, but in that of any woman at the appropriate time of her menstrual cycle. Edwards cautiously limits the application of his developing techniques to the provision of a healthy embryo for a woman whose oviducts are blocked and prevent descent of the egg. It should be obvious that the technique can be quickly and widely extended. The embryos produced in the laboratory might come from selected genotypes, both male and female. Preservation of spermatozoa in deep frozen condition can permit a high degree of selectivity among the sperm donors, who so far have been limited to the husbands of the women donors of the oocytes. Sex determination of the embryos is possible before implantation; and embryos with abnormal chromosome constitutions can be discarded. By checking the sperm and egg donors with a battery of biochemical tests, matching of carriers of the same defective gene can be avoided or the defective embryos can themselves be detected and discarded. By preserving the reproductive cells obtained from young persons under conditions which minimize mutation, the same individuals may have offspring at a relatively advanced age without incurring the higher probability of adverse gene and chromosome defects that normally increases with age. In the future age of man it will become possible for every person to procreate with the assurance that the child, either one's own or one prenatally adopted, has a sound heritage, capable of fully utilizing the opportunities provided by society for optimal development.

We do not know enough about the more complex aspects of human nature, such as intelligence and personality, to exercise strong selection wisely. However, if every couple were permitted to have only two children, or to exceed that number only upon special evidence that the first two are physically and mentally sound, a mild eugenic practice would be introduced that is probably all mankind is prepared to accept at this time. Let us insist, in any case, upon avoiding any measures that might decrease the extent of human genetic diversity. The intelligence of man is an evolutionary product of natural selection for adaptability to great variation of surroundings, to tremendous vicissitudes of experience, as Bergson concluded 60 years ago. That intelligence depends upon a genetic pool in the populations of man to which many genes contribute, in ever-shifting combinations produced by the patterns of heredity and reproduction. A pure line, no matter how perfectly selected for special excellence under given conditions, could be well adapted only to particular conditions. Unless man's power produced a world not only stable but unvarying and unvaried, the maintenance of human diversity should be a paramount aim. As long as our brave new world presents an abundance of choices and as long as we have freedom to choose, so long will human intelligence based upon genetic diversity remain a primary requirement.

Intelligence without integrity will fail. Roderick Seidenberg, in a later book, *Anatomy of the Future*, puts it well: ". . . the means rather than the ends of life are in our command." Man requires a challenge and a quest if he is to avoid boredom. The Golden Age toward which we move will soon look tawdry if we no longer see endless horizons. We must, then, seek a change within man himself. As he acquires more fully the power to control his own genotype and to direct the course of his own evolution, he must produce a Man who can transcend his present nature. The Erect Ape-Man had little vision of the power that his 20th-century descendant would wield in really so short a span of evolutionary time. For the Erect Ape-Man a club was the acme of power. Even so, if Man can avoid the ultimate follies which our present powers have bestowed upon us, and can survive a few centuries more, we today can little perceive what he may be. Perhaps the Golden Age of no progress will be but a passing phase and history may resume. We can only hope.

Index